The Mystery of Sleep

The
Mystery
of Sleep

Why a Good Night's Rest Is
Vital to a Better, Healthier Life

MEIR KRYGER, M.D.

Yale
UNIVERSITY PRESS

New Haven and London

The information and suggestions contained in this book are not intended to replace the services of your physician or caregiver. Because each person and each medical situation is unique, you should consult your own physician to get answers to your personal questions, to evaluate any symptoms you may have, or to receive suggestions for appropriate medications.

The author has attempted to make this book as accurate and up-to-date as possible, but it may nevertheless contain errors, omissions, or material that is out of date at the time you read it. Neither the author nor the publisher has any legal responsibility or liability for errors, omissions, out-of-date material, or the reader's application of the medical information or advice contained in this book.

Published with assistance from the foundation established in memory of Amasa Stone Mather of the Class of 1907, Yale College.

Yale University Press books may be purchased in quantity for educational, business, or promotional use. For information, please e-mail sales.press@yale.edu (U.S. office) or sales@yaleup.co.uk (U.K. office).

Designed by Sonia L. Shannon.
Set in Electra type by Integrated Publishing Solutions.
Printed in the United States of America.

Library of Congress Control Number: 2016952688
ISBN 978-0-300-22408-5 (hardcover : alk. paper)

A catalogue record for this book is available from the British Library.

This paper meets the requirements of ANSI/NISO Z39.48–1992 (Permanence of Paper).

10 9 8 7 6 5 4 3 2 1

This book is dedicated to my wife, Barbara, and my children,
Shelley and Jason, Michael and Emily, and Steven

Sleep that knits up the ravell'd sleave of care,
The death of each day's life, sore labour's bath,
Balm of hurt minds, great nature's second course,
Chief nourisher in life's feast.

—SHAKESPEARE, *MACBETH* (2.2.48–51)

CONTENTS

PREFACE

SLEEP HAS ALWAYS BEEN A mystery to me. Almost two decades ago, a woman with narcolepsy and I were interviewed on a live radio program about the impact of sleep disorders on daily life. A couple of hours later, when I was back in my office, I received a phone call from a literary agent, who asked me how sleep disorders affected women. As we chatted, it became obvious that the general public did not appreciate that women suffer from many sleep problems that men do not, and that because of the variety of roles they play in society, sleep disorders affect women differently from the way they do men. I had written or edited several books about sleep for doctors. It was time for a book for women; this was born in 2004. Now, in this new book, I explore the impact of sleep disorders on the entire family and society.

Although sleep disorders are common and there has been an upsurge of interest in them in the past few years, millions of people still suffer from the effects of undiagnosed sleep problems. The National Sleep Foundation estimates that up to 47 million American adults may be putting themselves at risk for injury and health and behavior problems because they are not getting enough sleep. The National Institutes of Health estimates that between 50 and 70 million Americans of all ages have sleep-related problems, including insomnia and sleep apnea.

In the mid-1970s, when I reported one of the first cases in North America of what eventually became known as sleep apnea, I thought I was describing a rare condition. In the next two decades it became apparent that sleep apnea was as common as asthma, and it affects millions of people worldwide. The condition was not new, but people who had it were being misdiagnosed and treated for the wrong conditions. Sleep apnea and other sleep-breathing disorders were seen as problems of overweight men who snore and were thought to be rare or even nonexistent in women. Until 1993 few doctors even looked for sleep apnea among women in the general population. It has since become clear that at least 2 percent of all adult women have sleep apnea. Just as men who had sleep apnea went undiagnosed for many years, until recently women with sleep apnea were misdiagnosed and often treated for other conditions such as depression.

More attention must be given to sleep disorders, which can cause misery and even death. Treatment is often delayed because a patient's sleep problems are not recognized. People with narcolepsy, for instance, on average wait fifteen years for a correct diagnosis, often after consulting many doctors. Yet this incurable condition has serious consequences. How can a person who cannot stay awake take care of children or have a career? One of my patients was referred to me only after she had fallen asleep while driving and crashed. Her two-year-old daughter was killed.

Sleep apnea has also been linked to heart disease. People with sleep apnea repeatedly stop breathing, which reduces blood oxygen, causing high blood pressure and stressing the cardiovascular system, potentially leading to heart disease or stroke, and possibly hastening cognitive decline in the elderly. Most people do not appreciate how deadly certain sleep disorders can be, but it is vitally important that sufferers of sleep disorders get a correct diagnosis and the proper treatment.

Over the past forty years, I have treated more than thirty thousand patients with sleep problems and have seen firsthand the disastrous consequences of years of misdiagnosis and lack of treatment. I wrote this book to help raise awareness of the effects of sleep disorders on people's daily (and nightly) lives. Sleep problems can occur at any stage of life. This book is intended to provide the tools people need to recognize and understand their own sleep disorders and those of family members and to help them find treatment. My goal is to educate and empower people, to enable them to stay awake and alert so they can enjoy life to the fullest.

ACKNOWLEDGMENTS

A SPECIAL THANKS TO Dr. Norah Vincent, a colleague and clinical psychologist, who provided the content on the behavioral treatment of insomnia. Swen Salgadoe, a former Northwest flight attendant whom I am proud to call my niece, is an expert in overcoming jet lag and supplied an insider's view on how to do it.

Thank you to Jean Thomson Black, Mary Pasti, Samantha Ostrowski, Susan Laity, and the staff of Yale University Press for help and encouragement in all the steps in the production of this work.

Thank you as well to my patients, who taught me how sleep disorders affected their lives.

Finally a huge hug, kiss, and so forth to my wife, Barbara, who reviewed and edited the text and often put up with my laptop on the corner of the kitchen table while I was writing this.

·✦·

PART ONE

A Good Night's Sleep

1

Why Do We Sleep?

THE MYSTERY. Why do all forms of life, from plants, insects, sea creatures, amphibians, and birds to mammals, need rest or sleep?

~~~~~~~~~~~~~~~~~~~~~

Our Sleepy Lives

It is 3 A.M. You awaken sweating, your heart pounding. Your mind is racing, reminding you that you have to get up early to drop the kids off at daycare, then dash to work for an important meeting. Obsessed with your personal and work issues, you eventually fall back asleep, only to be dragged awake by the alarm clock. You are cranky and yell at the kids, trying to rush everyone out of the house. By the time you get to the meeting, you have trouble paying attention; you can neither concentrate nor listen to others. You gulp down a second cup of coffee.

As you drive down the highway you feel an uncontrollable urge to shut your eyes. You turn on the radio, pop a candy into your mouth, and start to sing along with the radio. You roll down the window. Maybe you slap your face. But nothing seems to be working. You cannot overcome the urge to fall asleep.

If you have ever been affected by severe sleepiness, you are not alone. Every year, U.S. businesses lose an estimated $18 billion in productivity and injuries because of daytime sleepiness, while sleepy drivers cause an estimated 20 percent of car accidents, as high as 1.2 million crashes, resulting in thousands of deaths and injuries and billions of dollars in property damage. Research has also shown important links between the amount of sleep an individual gets and risk of health problems: reduced sleep may lead to obesity; women who sleep much more or much less than the average are at increased risk of disease. A study of 71,000 nurses published in 2003 showed that those sleeping five hours or less had a 45 percent greater risk of developing heart disease after ten years than those sleeping eight hours. Those sleeping nine to eleven hours increased their risk by 38 percent. Sleep disorders, ranging from mild insomnia to sleep apnea to narcolepsy, affect an estimated 50 to 70 million Americans.

Why do our bodies need sleep? How does sleep deprivation affect us, and how can we recognize its signs? In this book I address these questions and others. Sleep medicine is such a new field that even sleep specialists are still learning about it, and many mysteries remain about the nature of sleep and its disorders. I tell readers how to recognize the symptoms of sleep problems they or their family might be suffering from and how they can manage these problems so that they get enough of the sleep they need.

What Is Sleep?

All forms of life have periods of activity alternating with periods of inactivity, taking the form of either rest or sleep. The amount of sleep each lifeform needs, and its sleep schedule, is controlled by its genetic blueprint—and we humans share some of our sleep-controlling genes with flies!

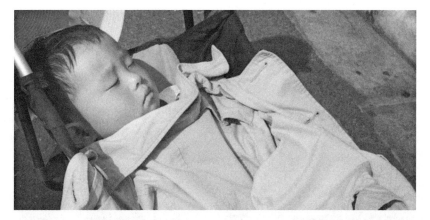

Yet despite the thousands of experiments scientists have performed to study sleep, no one has been able to declare with certainty why all lifeforms need sleep; we know only that when animals are prevented from sleeping they eventually die. It is possible that sleep serves different functions for different species. Most animals, for instance, spend their waking time looking for food, then retreat to a safe place to sleep and hide from predators. (Animals near the top of their food chain, such as lions, seem to sleep whenever and wherever they want.)

Some of the reasons suggested for why humans sleep include removal from the brain of waste material produced by brain cells, conservation of energy, the restoration of important bodily functions, and the repair of damaged tissues. Certain hormones, for example, are secreted mainly during sleep. The one thing scientists are certain of is that if people do not get enough sleep, their brains will not work properly; they will feel rotten, and they might find themselves unable to perform complex tasks. No one wants a sleep-deprived pilot flying a commercial jet across the Pacific. And even the medical profession now admits that doctors who are sleep deprived (a condition that used to be the norm among residents, who were on continuous duty for up to thirty-six hours and a hundred hours a week) are a hazard to their patients and their own health.

We can conclude that sleep serves many important functions, and different types of sleep serve different needs. For instance, scientists theorize that slow-wave sleep makes us feel refreshed upon awakening, while rapid eye movement (REM) sleep is connected to our ability to store memories in our nervous system. The right amount of sleep contributes not only to our ability to function but to how we feel. Sleep may in fact be, as Macbeth describes it, the "chief nourisher in life's feast."

Since ancient times scientists, philosophers, artists, and writers have been fascinated by sleep and dreams. Up to the nineteenth century, many believed that sleep was a form of death: as Dr. Robert MacNish wrote in *Philosophy of Sleep* (1830): "Sleep is the intermediate state between wakefulness and death: wakefulness is regarded as the active state of all the animal and intellectual functions and death as that of their total suspension."

In 1875 an English doctor, Richard Caton, reported that he could measure electrical activity in the brains of animals. It was not until the twentieth century that scientists discovered that the human brain was active during sleep and that they could measure its electrical activity. In 1928 a German doctor, Hans Berger, successfully recorded the electrical activity of a sleeping brain using electrodes placed on the scalp. This was the first EEG (electroencephalogram). Soon techniques were devised to measure the millionths of volts of electrical energy put out by the human brain during sleep. In 1953 Nathaniel Kleitman of the University of Chicago and his student Eugene Aserinsky used EEGs to measure electrical activity in sleeping infants, while also measuring their eye movements, becoming the first to describe REM sleep. Scientists then realized that there are three states of consciousness in all mammals: non-REM sleep, REM sleep, and wakefulness.

It was soon discovered that REM sleep is the time when sleepers are most likely to experience vivid dreams. Scientists then learned that there were sev-

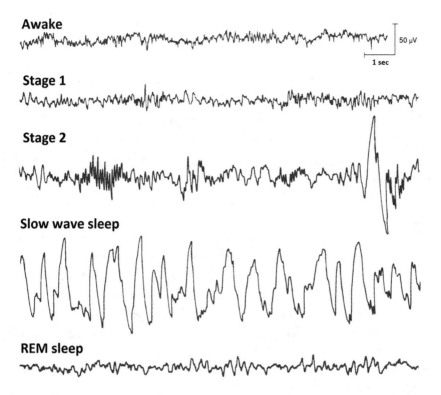

Awake

50 µV

1 sec

Stage 1

Stage 2

Slow wave sleep

REM sleep

Brain waves

eral types or stages of sleep. In 1968, at a meeting of an international group of sleep scientists, a more detailed picture of the brain's electrical activity during sleep emerged. Using brain waves and other measures, they initially segregated non-REM sleep into four stages. Typically, as a sleeper went from stage 1 to stage 4, the brain waves were shown to move progressively more slowly and their size became increasingly bigger as the sleep became deeper. Stages 3 and 4 are now considered together as slow wave or deep sleep.

Paradoxically, a normal feature of sleep is that we experience brief awakenings (also called arousals), each lasting only seconds, while we sleep. These arousals occur from the time we are born, and healthy sleepers might experience about five of these awakenings an hour, although they will not remember them. Scientists believe that arousals represent a response to a sensation or a protection from danger: a newborn, for instance, must be able to arouse if its breathing passage is blocked by bedding, so that it can move into a position where its breathing can start again. An adult who stops breathing during sleep must also arouse to start breathing again. People with some of the diseases described in this book wake up ten times more often than healthy sleepers. The many awakenings reduce the amount of time these sleepers spend in the various sleep stages and prevent them from experiencing the amount of continuous high-quality sleep their bodies and brains need. The result is often daytime sleepiness and difficulty performing daily activities.

REM Sleep

Although scientists have uncovered a great deal about rapid eye movement sleep, there is still much to learn. As illustrated in the figure above, brain activity during REM sleep is similar to that seen in wakefulness. Furthermore, the brain cells use a great deal of energy, another indication of brain activity, which has led some scientists to theorize that REM sleep might play an important role in learning, laying down the networks in the brain that allow us both to learn new things and to remember what we've learned.

Curiously, at the same time that our brains are experiencing this activity during REM sleep, almost all of our muscles are paralyzed, with the exception of the major breathing muscle, the diaphragm, and certain sphincters at the top and bottom of our gastrointestinal tract.

In spite of this general paralysis, cells in the pons, a vital part of the brain that is also involved in controlling breathing, become active and create electrical storms, with electrical impulses that work their way through the central nervous system until they reach the part of the brain that controls eye movements, setting off the rapid eye movements characteristic of this state of sleep. When these impulses pass through the parts of the nervous system that control breathing and the cardiovascular system, they can create dangerous irregularities in the pattern of breathing as well as in heart rate and rhythm and blood pressure. Sleep apnea in women, for example, might only occur during REM sleep.

Perhaps the most intriguing—and mysterious—feature of REM sleep is that most vivid dreaming occurs during this enigmatic state. All humans with intact brains dream, usually three to five times a night. Although scholars and thinkers ranging from the indigenous Australians to the seventh-century

B.C.E. Assyrian king Ashurbanipal to Sigmund Freud have ascribed meanings to dreams, we still know very little about their origins or function.

In about 400 C.E., a Roman by the name of Macrobius wrote a treatise in which he characterized five different types of dreams: the mysterious dream that requires interpretation (as Freud later did); the prophetic dream that comes true (as happened in the biblical story of Jacob's ladder dream); dreams in which authority figures direct the dreamer (such as the saints directing Joan of Arc in her visions); dreams related to nightmares (as are found with post-traumatic stress); and nightmares in which there is contact with an apparition (such as Scrooge's encounters in Charles Dickens's *Christmas Carol*).

We know, for example, that all animals apparently experience REM sleep, but do they dream? When a puppy thrashes about or barks or appears to run during sleep, is it dreaming? Newborns spend half their sleep time in REM sleep. Are they dreaming?

Every time a male dreams, he has an erection, and every time a female dreams, the blood vessels of her vagina become engorged. Yet it appears that sexual thoughts before sleeping or explicitly sexual dreams do not cause these events; they are the result of the state of dreaming itself.

Some people react physically to what they are dreaming, instead of being paralyzed.

We do not know why some people who have suffered psychological trauma awaken for years with dreams that replay that trauma.

How Much Sleep Do We Need?

How much we sleep, when we sleep, and how much deep sleep and dreaming sleep we need vary with age. Seven to nine hours of sleep, which is adequate for most adults, would not be enough for the average nine-year-old. The table below shows the range of sleep required at different ages. For all age groups, however, the amount of sleep a person needs is an individual characteristic just as his or her height is.

The amount of sleep an individual needs each night is however much will enable him or her to remain wide awake and alert the next day, and it varies over the individual's lifespan, decreasing with age. Babies spend most of their time sleeping (although it may not feel that way to their sleep-deprived parents), and they do not follow a pattern: for the first few months of life, in-

Sleep Needs, by Age

Age	Hours asleep of 24	Hours spent napping
Birth to 2 months	10.5 to 18	5 to 10
2 months to 12 months	14 to 15	2.5 to 5
12 months to 18 months	13 to 15	2 to 3
18 months to 3 years	12 to 14	1.5 to 2.5
3 years to 5 years	11 to 13	0 to 2.5
5 years to 12 years	9 to 11	0
13 years to 20 years	8 to 10	0
Over 20	7 to 9	0

Note: A newborn will sleep at any time during the day or night. Excessive napping after age 5 may indicate a sleep problem.

fants might sleep at any time during the twenty-four-hour day. Mercifully for the parents, they soon start to spend periods in sleep mostly at night. Infants and toddlers nap. By the time children go to school, most will no longer nap.

The amount of REM sleep also decreases with age. Newborns spend roughly half their sleep time in REM sleep. In adults the amount of time spent in REM sleep decreases to 20–25 percent. The amount of slow-wave sleep is also much higher in children, since this is the sleep state during which most of the human growth hormone is secreted; the time most people spend in slow-wave sleep decreases with aging. Some elderly people have no slow-wave sleep.

As will be detailed later in this book, few of us get the amount of sleep we need. Today's world and its demands on our time have eaten away at our sleep time. In teenagers the deficit can be more than two hours every night. The average American sleeps fewer than seven hours a night.

Teenagers, in particular, tend to develop bad sleep habits. They go to sleep later and awaken later than they did during childhood years. On school nights, it may take them several hours to fall asleep, and they are unable to get the eight to ten hours of sleep they need to be alert throughout the day. Parents may find themselves dragging an unwilling and unresponsive teenager out of bed, with the result that for the first few hours of the school day, the teenager might seem to be in a daze or might actually be asleep and consequently is likely to perform poorly. On weekends, these teenagers generally sleep until noon or later. For them to try to be alert at 9:00 A.M. would be like the average adult trying to be alert at 2:00 or 3:00 in the morning. It simply does not work. In the afternoon and evening, these adolescents frequently get a second wind.

Older persons, especially after retirement, might start to nap again. Whether reduced nighttime sleep in the elderly is a consequence of the day-time naps is not clear. Many elderly people sleep poorly not just because of their age but also because of medical conditions, medications, pain, sensitivity to their environment, or disruptions and changes in their sleep pattern. A survey of sleep in the older population, released in 2003, showed that older people without medical problems usually sleep the normal amount of time for their age group.

How the Brain Controls Our Sleep

Research has been done to find the chemicals and brain cells and pathways that control when humans sleep and when they wake up. Many brain structures are involved, as can be seen in the diagram of the brain centers (below) that control sleeping and waking. The darker-shaded boxes indicate structures involved in sleep, the lighter-shaded boxes, structures involved in wakefulness. However, to understand what starts and stops sleep we need to understand two concepts: the wake gauge and the body clock.

THE WAKE GAUGE

The fuel gauge in a car indicates when the driver needs to refuel the car. The wake gauge indicates when it is time to sleep. Once the average adult has been awake for about fourteen hours he or she starts to become sleepy, and the sleepiness becomes greater at sixteen hours, and much greater at eighteen hours, when it becomes difficult to stay awake. The wake gauge in the brain measures the amount of a chemical called adenosine, which is involved in the transfer of energy in the body. The longer a brain is active and using energy, the greater the concentration of adenosine. Adenosine acts to promote sleep

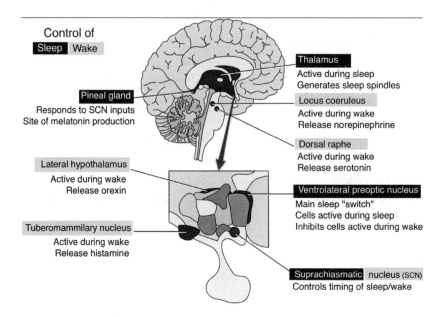

Brain Centers That Control Sleeping and Waking Functions

and suppress wakefulness. Caffeine counteracts the effect of adenosine, which is how it keeps people awake.

THE BODY CLOCK

In addition to influencing the amount of sleep the body gets, the brain helps control the time when people sleep. How does the body know when it is time to go to sleep and when to wake up? A collection of cells in the suprachias-matic nucleus (SCN) of the brain has the ability to keep time and monitor ithe sleep-wake cycle. A hormone called melatonin is secreted by a tiny gland in the brain called the pineal gland when it starts to become dark, at dusk. The SCN cells control not only the times when individuals are sleepy or alert; they also control the function of many other systems in the body. The word *many*, in fact, may be an understatement. Most of the systems in the body have a pattern that varies over a twenty-four-hour period. This is true of the secretion of a variety of hormones, of blood pressure, of heart rate, and of some other functions in the body.

This natural, internal rhythm is called the circadian rhythm. The word derives from *circa*, "about," and *diem*, "day." The circadian rhythm changes the way many systems in the body work over the twenty-four-hour day so that the function of the systems matches what the body needs. Indeed, it has been shown that tissues in the body far away from the brain (such as the liver and kidneys) also have circadian clocks that are synchronized to the master clock in the brain. As a result, humans generally do not experience hunger or a need to go to the bathroom at night. People who have traveled across time zones know how discombobulated or out of sync they can feel because their body clock is not on the same time as the place they happen to be.

For many years, scientists wondered how the brain knew what time it was. How could it tell whether it was morning and time to wake up? Suppose we were living on a planet that was rotating around a sun every thirty hours instead of twenty-four hours—how would the body clock adjust? Studies conducted at Harvard University, the University of Pittsburgh, and other academic centers revealed that the light could reset the body clock. Exposure to morning sunlight was responsible for synchronizing the body's circadian clock, in humans and other animals (and would do so for a thirty-hour day on another planet). Light enters the eye and hits the retina, stimulating specialized cells. The visual information travels from these cells along nerves to the suprachiasmatic nucleus, where the cells that control circadian rhythms reside. These cells are located above the optic chiasm, where visual information crosses over from one side of the nervous system to the other on its way to the part of the brain that processes vision. The SCN uses the information from the retina to tell the brain that it is morning; this in turn synchronizes the SCN cells. People who are blind because of problems between the eye and the optic chiasm can have serious difficulty in synchronizing their body clocks, and often have severe sleep problems as a result. On the other hand, people whose blindness results from problems in the visual cortex, the part of the brain that processes the information, might have a normal circadian system.

Circadian systems appear to be present not only in higher life forms, but also, for example, in plants. The chemical melatonin that is produced in the brain at dusk is also found in forms of life ranging from animals to insects to jellyfish to bacteria and even to plants. The first demonstration of a circadian rhythm was given by Jean-Jacques d'Ortous de Mairan, a Swiss scientist who set up an experiment using a mimosa plant that always opened its leaves at a certain time when it was sunny. He put the plant into a box, where it received no exposure to light, and the leaves still opened at the same time. This plant was able to keep track of time.

Life forms have existed on earth for more than a billion years. Mammals have inhabited the planet for about 200 million years. The dinosaurs disappeared about 65 million years ago. The earliest ancestors of humans appeared about 7 million years ago, and humans that resembled us arrived only two hundred thousand years ago. Humans evolved as diurnal beings: active during the day and sleeping when it is dark. Many other animals are nocturnal, sleeping during the day and active at night. It is likely that until about 150 years ago, with the arrival of artificial light, humans' sleeping habits did not change much. Prehistoric people living in caves probably slept about two hours more than the average adult sleeps now. But people have not always slept in one continuous period at night. Long before the invention of the lightbulb, many people had two sleep periods a night. The first sleep started about two hours after sunset and would last three to four hours. The second sleep, also about four hours long, would start one to three hours later. Between the two sleeps the person might pray, read, or participate in sexual activity. The middle of the night was a literal "midnight" between these periods. In the past 150 years there have been dramatic changes in the quantity and pattern of human sleep. People sleep less, and they no longer sleep exclusively in the hours between dusk and dawn.

(A fascinating side note on our forebears: Although most probably slept on flat surfaces, ranging from the hard ground to soft mattresses, at various times and places sleeping sitting up was the norm. You can still see ancient buildings in the city of Bergen, Norway, in which the occupants slept in enclosed cubby holes that forced them to sleep sitting up. In Antwerp, one of the highlights of a visit to the home of Peter Paul Rubens is the short bed in which he slept sitting up.)

There are many time-related rhythms in biology, not all of which are understood. Some biological rhythms can be measured in seconds (for example, breathing and heartbeat), while others have much longer cycles. Although the cycle we know the most about is the circadian rhythm controlled by the body clock in our brains, seasons of the year seem to play an important role in when mammals can become sexually active, when they become pregnant, and when they give birth. One of the most mysterious cycles in biology is the menstrual cycle in women, which is reviewed in Chapter 3. This cycle is about the same length as the lunar cycle, and the timing system that results in the cycle, which averages twenty-eight days, is unknown.

How to Recognize When You Have Slept Well

No matter what your circadian rhythm or body clock is, or what position you sleep in, you should be able to recognize a healthy sleep pattern. After a good night of sleep you should feel wide awake and alert shortly after waking up, and remain so for the rest of the day. Your mood is generally good, and you feel no need for a nap. A good night of sleep is a matter of both quantity (it should be the right amount for your age group) and quality (it should be uninterrupted and consist of the right amount of each stage of sleep).

You should not wake up feeling as though you have not slept. You should not feel as though you won't be able to function until you have had one or more cups of coffee. Struggling to stay awake while driving or falling asleep or feeling uncontrollably fidgety at movies, public meetings, or even in front of the television or computer screen are all signs that you may be sleep deprived. You should not feel as though you are about to fall asleep when reading.

If you do experience these symptoms, it indicates that the amount or the quality of your sleep is inadequate to keep you optimally awake and alert. You probably have a sleep problem if you are sleepy in the morning, feel tired all day, fall asleep when you don't want to, need to nap, and are irritable and moody when you awaken. It is important to note that boredom does not cause sleepiness. Boredom simply gives the sleepy person the excuse to nod off.

Other symptoms, which will be discussed later in the book, might indicate that you have a medical problem. These include waking up with heartburn, chest pain, shortness of breath, or an unusually fast or slow heartbeat. Waking up with a headache more than just occasionally or having to make frequent trips to the bathroom at night could also be signs of a medical problem. You should not wake up unable to move or with severe sweating. Nor should you be thrashing around in a way that could injure yourself or others while you are sleeping. Bed partners should not be telling you that you stop breathing during sleep and that it is scary to watch you sleep. If you have any of these symptoms, they probably relate to a medical problem that should be investigated.

You should also see a doctor if you are an adult who sleeps more than ten or less than five hours a night. Research has shown repeatedly that people who consistently sleep too much (more than ten hours a night) or too little (fewer than five hours a night) have a higher death rate than those who sleep the appropriate length of time. However, the key issue is not the length of sleep. Such abnormal amounts of sleep are rather a symptom of a sleep or medical disorder that may cause or result in death.

The bottom line is that if you are not wide awake and alert throughout the day, if you experience daytime sleepiness, or if you have any of the symptoms I just described, you may have a problem. Your sleep problem could affect you, your family, or the entire world. Consider the cases of some recent U.S. presidents.

They have perhaps the most grueling job in the world. No days off. Stress. Travel. Always on call. This is a recipe for sleep deprivation. Sleep deprivation takes its toll. Did lack of sleep almost change U.S. history?

There is widespread speculation that a sleep-deprived Barack Obama was falling asleep during his first debate with Mitt Romney in Denver during the 2012 election. There are several possible explanations for Obama's sleep problem. The debate preparation resulted in sleepless nights. In addition, the city where the debate took place, Denver, is a mile above sea level. Some people at this altitude develop an abnormal sleep breathing pattern (a variant of sleep apnea) which causes them to have short awakenings during the night, resulting in restless or non-refreshing sleep. Despite reports that suggest Obama is a night owl who goes to bed late and gets up early, apparently getting only four to five hours sleep and sometimes not even that, such sleep deprivation might have accounted for his poor showing in Denver. He needs to sleep more. He mentioned at a prayer meeting that after he leaves office, "I am going to take three, four months where I just sleep."

Ronald Reagan fell asleep during an audience with the pope. Bill Clinton had meetings that went through much of the night, as reported in *Newsweek*: "Over the years Clinton had tried to convince himself he could get by just fine on a few hours of sleep a night. Time and again, he proved himself wrong. Struggling to extricate himself from a previous scandal, Clinton once told a friend, 'Every important mistake I've made in my life, I've made because I was too tired.'" William Taft, who weighed as much as 350 pounds (there is to this day an extra-wide "Taft chair" in Woolsey Hall at Yale), had sleep apnea and was sleepy while in office. The most powerful people in the world are sometimes almost incapacitated by their sleep problems.

We Are Not All the Same, but We All Need a Good Night's Sleep

With all that said, some very high-functioning people sleep more than ten hours a night, while some high-functioning people sleep less than four. It is important to remember that data obtained from a population may not apply to an individual.

The importance, control, and complexity of sleep is something we have only recently begun to understand. A good night's sleep can make a person feel wonderful and be highly productive. A poor night's sleep or insufficient sleep leaves a person feeling exhausted and nonproductive; he or she may even pose a danger to others. Women are much more likely than men to have sleep problems. This is not just because some sleep disorders are more common in women but also because family responsibilities, menstruation, pregnancy, and menopause can disrupt normal sleep.

Often in a household the main caregiver does double duty. She or he often both works outside the home and acts as the primary caregiver, and this person is the first up in the morning and the last to go to bed at night. The primary caregiver often organizes and runs the household, prepares meals, and does the housecleaning. These caregivers need to be emotionally alert to the needs of spouses and children, to problems and issues in the family, and so on. If they lose sleep, not only they but the entire family can suffer. They can be cranky, irritable, and short-tempered.

For those performing these roles, a good night's sleep is essential. After a good night of sleep people start the day fresh, with enough energy to take them through the necessary tasks. For a healthy lifestyle, a good night's sleep is as important as healthy eating and adequate exercise. But not everyone recognizes sleep problems when they arise. In the following chapters I offer the information readers need to recognize when they have a problem, and what they can do about it.

2

Sleep Requirements
in the Life Stages

THE MYSTERY. Why do humans have different sleep needs at different stages of life? Unless they understand what to expect, they can place themselves at risk for sleep deprivation, with its potential to impair school and work performance, and to disrupt the family.

The Case of the Sleepy Teenager

I can always recognize a patient who doesn't want to be in the sleep clinic where I work. It was 8:00 A.M. The thin young man, about seventeen years old, was sitting slouched in the waiting room wearing a hoodie, sunglasses, and earbuds connected to a smartphone, sound asleep. Sitting next to him, looking very concerned, was a well-dressed woman, presumably his mother. Minutes later in the exam room his story came out, with his mother doing all the talking. The young man had been a straight-A student, and now he was failing high school. His teachers were worried that he might have attention deficit hyperactivity disorder (ADHD) or a disease because he continually daydreamed or fell asleep in his classes. His mother explained that it was a struggle every morning to get him out of bed; he could sleep through the ringing of two alarm clocks.

The whole time his mother was speaking, the young man sat texting. (I was amazed that he could see the screen since he still wore his sunglasses.) When she finished, I asked him about whether he snored, had vivid dreams as he was falling asleep, ever awoke paralyzed, or used street drugs, along with a number of other questions that would help me rule out various diseases. The answer was always one annoyed word: "No." I then asked him whether he thought he had a sleep problem. "No!" His mother looked angrily at him. I asked him to remove the hoodie and the sunglasses, so I could examine him. The high school football insignia on his T-shirt and bags under his eyes told me almost all I needed to know, and when he described his daily schedule, it confirmed my suspicions.

Sleep Needs of Children

From birth, every child has individual sleep needs and patterns. These patterns change with time, and the way the patterns change is also individual. If a couple's first baby is a great sleeper, this does not mean that future children will be great sleepers. Yet despite individual idiosyncrasies, children tend to have similar sleep needs and patterns at the same ages, and parents who understand what to expect and learn their child's sleep pattern can recognize when problems arise and when to seek help. In 2015 the National Sleep Foundation published a report (summarized in the figure below) listing recommended hours of sleep for nine age groups, from infants to the elderly.

THE FIRST YEAR

Sleep patterns change tremendously in the first few months of life. At first, neither the baby nor the parents sleep through the night and the parents experience sleep deprivation. This can be a difficult time for all. It is also the time when a small number of mothers may experience postpartum depres-

Hours of sleep

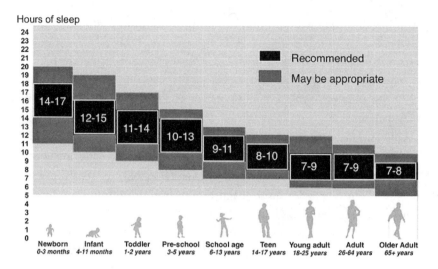

Hours of Sleep Needed a Night, by Age

sion, which can be very serious and often requires treatment. Most parents find relief only after the baby begins to sleep through the night.

Newborn babies have many sleep periods over the course of a twenty-four-hour day, most of which last between a half hour and three hours. As the newborn does not distinguish between day and night, it awakens many times throughout a typical night. By six weeks, nighttime sleep episodes have become longer, and a more regular sleep pattern usually begins. The baby will sleep a total of fourteen and a half hours on average, with the range being between ten and eighteen hours. Newborns spend about 50 percent of their sleep time in REM sleep, during which they twitch and grunt and boys may have erections. All of this is normal.

Parents who want to help their baby get the sleep it needs must learn the baby's language—that is, how he or she communicates signs of sleepiness. For instance, certain cries, fussiness, and rubbing of the eyes signal that the baby is sleepy. Parents should recognize that the baby is tired when they see those signs and put the child to bed. (This would also be a good time for the parents to sleep, unless there are other children who require attention.) Early on parents might rock the baby to sleep, but by the end of the second month, most babies should be learning to fall asleep on their own.

Between two and twelve months, the baby's nighttime sleep episodes

gradually become longer. By four to six months of age he or she will temporarily wake up during the night but should begin to fall back to sleep without
intervention from the parents. At two months a baby will take three or four
naps over the course of twenty-four hours. By the end of the first year, most
babies take two naps a day.

Babies at this age should be put to bed on their back when they are ready
to sleep. Do not put a baby to bed every time the baby yawns — not every yawn
means the baby is ready for sleep. Parents should learn when their baby will
be sleepy and follow that schedule so that they won't have to wait for the baby
to show signs of sleepiness before putting her or him to bed. After the baby is
in bed, parents should fight the impulse to check on the child every time she
or he awakens, as babies need to fall asleep on their own even in the middle
of the night. If the baby begins to associate falling asleep with being held or
rocked by a parent, he or she might not fall asleep without this assistance —
babies quickly learn that if they cry, the parent will come and rock them.
Parents, meanwhile, are thinking, "If I don't go, the baby will never fall asleep
and I'll never get to sleep. I'd better go." Parents and baby are now trapped in
a vicious circle. This is the time when it is natural for babies to learn to fall
asleep on their own; parents should help them achieve this (see Chapter 7).

Babies who sleep more than eighteen hours or fewer than ten, are excessively sleepy or nonresponsive while awake, or exhibit loud snoring or obstructed or stopped breathing during sleep may have apnea. Causes at this
age include various abnormalities of the anatomy of the breathing passage or
something as simple as enlarged tonsils. Parents who suspect their child has a
disorder should contact a doctor. If the baby turns blue, or gray, get medical
help immediately as this is a sign of low blood oxygen.

ONE TO THREE YEARS

At one year of age, children are still napping, but they do most of their sleeping at night. Naps are generally quite regular. Their sleep time will usually
be eleven to fourteen hours a day. By the age of three, they are taking shorter
naps. It is during these years that children need to learn good sleep habits.

FOUR TO THIRTEEN YEARS

By the time the child is five, he or she generally will not need a nap. Children
age four to thirteen need at least two to three hours' more sleep than adults.

Over the years, I have seen several families whose children were falling asleep in class; the parents could not understand why this was happening because the child was getting as much sleep as they were. But although seven or eight hours of sleep a night is fine for the parents, it is inadequate for a child. All parents and educators should understand that children, from birth through adolescence, need more than the average amount required by adults.

Several sleep problems that can begin when the child is younger can become important issues in the years before adolescence. These include sleepwalking and bedwetting (enuresis). As we shall see in Chapter 14, sleepwalking is common among children, and as among adults, it can be made worse by sleep deprivation. Sleepwalking may be present from the time a child can walk but often becomes an issue later, for example when the child is invited for a sleepover. Sleepwalking generally becomes less frequent as the child grows older. If the sleepwalking episodes do not involve dangerous activity such as going downstairs or leaving the house, treatment is usually not required. Similarly, bedwetting often becomes a problem between ages five and ten. Parents of bedwetters should consult a pediatrician.

ADOLESCENTS, FOURTEEN TO SEVENTEEN YEARS

Many adolescents go to bed late and wake up late. Before bed, many spend their time in activities such as talking on the phone, playing video games, surfing the internet, or texting that tend to stimulate the brain and make it harder to sleep. The light from the screens of electronic devices may disturb sleep in part by suppressing the production of melatonin.

As a result of these habits as well as the natural tendency of adolescents to go to sleep later in the evening during their teenage years, American children are severely sleep deprived. The 2006 Sleep in America poll found that between the sixth and twelfth grades, teenagers' sleep time drops from 8.2 to 6.9 hours. Adolescents should sleep more than 9 hours. By the twelfth grade 95 percent of American children are sleep deprived!

The 2011 Sleep in America poll found that texting had become pervasive among adolescents: 72 percent of children ages thirteen to eighteen brought their smartphones into their bedrooms, and 56 percent texted during the hour before they went to bed.

In most school districts the yellow school buses appear at a time the children should be sleeping. If the child's schedule results in his or her falling

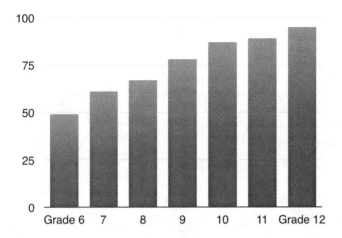

Percentage of Children Who Are Sleep Deprived, by Grade

asleep in class or performing poorly, parents should step in and make sure the child does not stay up late, does not participate in stimulating activities right before bedtime, and limits his or her caffeine intake. Parents who find it hard to keep their teenagers away from electronic screens might want to explore other options: some devices allow the user to change the colors of the screen to reduce blue, the wavelength of light that is the most disruptive.

Certain symptoms in sleep-deprived adolescents deserve special attention. Besides the sleep deprivation caused by a lifestyle that interferes with sleep, adolescents may find that their circadian clock changes, so that they are no longer on the same sleep schedule as adults, and this can cause problems (see Chapter 8). When the circadian clock runs late, teenagers go to bed late, have trouble getting out of bed to go to school, and sleep much later on weekends to catch up on sleep.

Teenagers who are extremely sleepy during the daytime and fall asleep at inappropriate times in spite of getting normal amounts of sleep may have a significant sleep disorder, and parents should consult a doctor. At this age, the possibilities include narcolepsy, which classically begins during the teenage years (see Chapter 13), sleep apnea (in a child who snores, the most common causes are enlarged tonsils and obesity; see Chapter 12), and some movement disorders (see Chapter 11). Teenagers who have an iron deficiency can also experience severe insomnia and daytime sleepiness, conditions that improve once the iron deficiency is treated.

Sleep Problems That Start in the Teenage Years

Sleep deprivation
Abnormal circadian clock
Sleep apnea
Narcolepsy
Movement disorder

Sleep Needs of Adults

YOUNG ADULTS, EIGHTEEN TO TWENTY-FIVE YEARS

Young adults, especially college students, are notoriously sleep deprived. They need seven to nine hours of sleep. They stay up late because of their lifestyle (spending time on social media, partying, studying) or because their circadian clock is running late (see Chapter 8) and may have to get up early to go to class. They pull all-nighters to finish projects and prepare for tests. In college classes, especially when there are hundreds of students in the room, as soon as the lights dim, many start to nod off. Students who learn about the importance of sleep and practice healthy sleep habits even during the college years can significantly improve their ability to have a healthy and productive lifestyle. Their success in life depends on these years.

ADULTS, TWENTY-SIX TO SIXTY-FOUR YEARS

Adults tend to need less sleep than children, teenagers, or young adults. But they do need on the average seven to nine hours a day. Generally the amount of sleep an adult gets is affected by his or her lifestyle, including family, social, and work obligations. Commuting, nontraditional work hours, and late-night time spent on social media chip away at sleep time. The quality and quantity of sleep can be affected by the development of diseases. For example, sleep apnea, discussed later in the book, first becomes symptomatic in people in their forties or fifties.

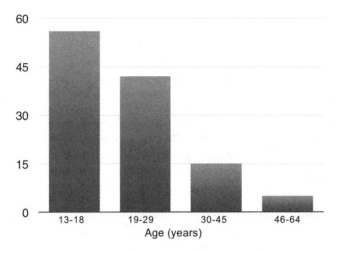

Percentage of Individuals Texting at Bedtime, by Age

OLDER ADULTS, SIXTY-FIVE YEARS AND UP

The majority of people in this age group are women, who on average live longer than men. The U.S. Census reported in 2013 that for every hundred women older than age sixty-five there were only seventy-eight men of the same age; for every hundred women age eighty-five or older there were only fifty-one men of the same age. It has been estimated that by 2060 the United States will have a population of 417 million. Of the total it is estimated that there will be 53 million women and 45 million men above age sixty-five. About a quarter of the entire population will be people over sixty-five.

But measuring age only in terms of years can be misleading. When I was a medical student, the average man was dead by the age of sixty-nine, and the average woman by age seventy-four. A person over sixty-five who became critically ill would not even be admitted into the intensive care unit because hospitals were unwilling to expend their resources on patients who were likely to die. Few Americans lived into their eighties, nineties, or hundreds. Today, in spite of air and water pollution, exposure to toxins in food, climate change, and other ills endangering life on our planet, people in the developed world are living longer and are healthier than at any time in the past. We have therefore had to rethink many of our guidelines with regard to best practices for "older" people.

Most important, we need to recognize that aging—deterioration in function related to time—varies from individual to individual. Within any given person, organ systems will age at different rates. Joints and muscles might age more quickly than the mind. Some people in their nineties use computers to trade stocks on the internet, and read several newspapers a day, yet have trouble climbing stairs or going for a walk. For other people, intellectual abilities begin to plummet in their early sixties while their bodies remain active. Some people contract diseases that seem to accelerate the aging process, such as diabetes.

Healthy older people can have normal sleep patterns—they fall asleep quickly, sleep through the night, and are wide awake and alert the next day. A 2003 poll concerning sleep in older people found that many are sleeping better and longer than younger people, and the better the health of older people, the better they slept. Like younger adults, older people need seven to eight hours of sleep every twenty-four hours, but some will start to take naps in the afternoon once they retire from work. They might then have trouble falling asleep at night or they might find themselves awakening very early in the morning. Some of my patients complain about these very early awakenings and ask what they can do to stay asleep until the time they want to wake up. They are surprised to hear that early awakenings need not be a problem if they are alert and functioning during the daytime. They can stay on their current sleep pattern, or, if they would rather change their nighttime sleep, they can try omitting the nap.

Napping can have risks, however; some recent studies have shown that there is an association between napping and long sleep duration with increased heart disease and diabetes. In fact, a tendency to fall asleep in the daytime or at inappropriate times and places can be an important clue that there is something abnormal about a person's sleep. An eighty-five-year-old woman brought her eighty-nine-year-old husband to see me. He still had his driver's license, but she was increasingly terrified to drive with him because he could not stay awake while driving. He had to take a fifteen-minute nap halfway through a thirty-minute drive from home to their country club. His excessive daytime sleepiness turned out to be a sign of sleep apnea.

Napping is particularly common in nursing homes. If you visit a nursing home during the daytime, you will often find lines of wheelchair-bound people who have nodded off. The residents of these homes seem to spend a great deal of time asleep. This may be due in part to the fact that they are not provided with enough exposure to natural sunlight, or the rooms may be poorly lit, which causes their circadian clocks to stop working.

Other abnormal sleep patterns may result from a variety of medical conditions, including high blood pressure, heart disease, diabetes, cancer, and depression; some people have one or more of these conditions. The more conditions a person has, the poorer the quality of his or her sleep. For example, people with four or more of these conditions are five times more likely to complain of daytime sleepiness than healthy people are. These diseases are covered in greater detail in Chapter 15. Women become menopausal in their late forties and early fifties, which causes sleep problems. They may also develop medical conditions that affect their sleep. I discuss these issues more fully in Chapter 5.

In addition to the common medical conditions of older people that can lead to sleep problems, many of the drugs used to treat those medical conditions can cause sleep problems. (In Chapter 17 I discuss which drugs cause sleep problems.) Older people are more sensitive to the main effects and side effects of medications than younger people, and both the health care provider and the patient should be informed about the side effects of every medication being considered.

It is also important for older people to be instructed about both the dosage and the regime of any medications prescribed. Sometimes aging patients' memories are not capable of tracking all this information, so they should make sure that it is written down and accessible. A particular worry is patients who continue to take old prescriptions because the bottles are still around, while also taking updated prescriptions of the same drug. Since many drugs have sleepiness as a side effect, such an unintended overdose of prescribed drugs can cause dangerous sleepiness or even coma.

Back to the Sleepy Teenager

The young man did not have a disease; he simply was not getting enough sleep. When I asked him to describe his schedule over a twenty-four-hour day, he told me that his mother woke him up every school day at 6:30 A.M. He would shower and dress quickly, but he never had time for breakfast before the school bus came at 7:00. School started at 7:30 and finished in the afternoon at 2:30. Then came football practice, from 3:00 to 5:00. His mother would pick him up from practice, and dinner was at 7:00. He often nodded off before dinner, then spent the rest of the evening rushing through homework on a big computer. At 11:00 P.M. he would start preparing for bed, but usually would not be in bed until midnight—and would fall asleep every night holding his smartphone, on which he had been texting. On weekends he would sleep until noon. Like most teenagers, this young man needed nine to ten hours of sleep a night, but his schedule was allowing him to sleep just over six hours. Because his mother, a mature adult, did not need as much sleep as her son, she did not realize that six and a half hours of sleep was simply too little for this stage of his life.

The solution to his problem turned out to be simple: no electronics after 10:00 P.M. and in bed by 11:00. The new schedule did not solve all his problems in school, but it helped. He was more alert throughout the day and his grades soon improved.

At different stages of life we need different amounts of sleep and follow different sleep patterns. Babies need to sleep most of the time, but parents can help them develop sleeping schedules that minimize their own loss of sleep. Teenagers who need to get up for the school bus must be ready for bed earlier than their parents. Older people might begin napping. At any age, failure to get the appropriate amount of sleep can lead to a variety of problems, from poor performance to serious illness. We all need to know how much sleep we need at each age, and this is something that parents, in particular, need to learn so they can help their children have healthy sleeping and waking lives.

3
The Reproductive Years

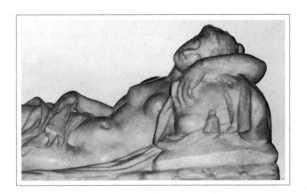

THE MYSTERY. Changes in sex hormonal function during the menstrual cycle, whether the cycle is regular or irregular, can have profound effects on a woman's sleep. Pregnancy can impact sleep. A common disease of the ovaries can cause sleep apnea.

~~~~~~~~~~~~~~~~~~

The Case of the Sleepy Woman with Irregular Periods

After almost five years of suffering from severe daytime sleepiness that made her unable to hold a job because she could not stay awake, a twenty-nine-year-old woman had reached the end of her rope. For months she had been awakening every morning with a headache. Her family doctor noted that these headaches, combined with the daytime sleepiness, were becoming progressively more serious and having a negative effect on the quality of her life. Her doctor also noted that she had been snoring loudly since she was a teenager.

She was referred to the sleep disorders center where I work for both her snoring and her sleepiness problem.

Her examination at the sleep center revealed she was overweight: at five foot five she was 160 pounds. What was more unusual about her was that she had more facial hair than is normal for a woman and very hairy arms and legs as well as hair on her chest between her breasts. The medical term for this pattern of excess hair is hirsutism. When I asked about her menstrual cycle she told me that her periods were very irregular; for many months she had not had a period at all. But what, she asked, could her menstrual cycle have to do with her sleep patterns? In her case, everything.

The Menstrual Cycle

In most females, starting about age twelve the menstrual cycle is monotonously regular, lasting about twenty-eight days, the same duration as the lunar cycle. Genes were identified in 2009 that play a role in controlling the age that menstruation begins. Certain genes control the body's circadian rhythm (see Chapter 8), and some of these genes are affected by the menstrual cycle, perhaps because of the accompanying profound changes in hormone levels. What starts each cycle remains a mystery. Research reported in 2015 showed that variations in the genes that control the circadian system can lead to irregular menstrual cycles.

According to the 2007 National Sleep Foundation poll of over a thousand women, 60 percent of American women get a good night's sleep only a few nights each week or less and 67 percent frequently experience a sleep problem. Additionally, 43 percent say that daytime sleepiness interferes with their daily activities. The sleep problems affect almost every aspect of their lives. They might leave home late or perform poorly at work, be too stressed or fatigued for sexual activity, or have little inclination to socialize. Sleep problems are experienced by women of all ages and increase in severity as they move through the different biological stages of their lives, which involve dramatic changes in the levels of reproductive hormones. The menstrual cycle is the most basic rhythm of a woman's life, yet millions of women have disturbed sleep because of menstrual symptoms.

Reproductive or sex hormones affect many organs of the body, including the brain: an abnormal amount and type of sex hormones can be the cause of serious medical and sleep problems. For example, women are more likely to develop symptoms related to depression at times when the levels of these hormones are increasing or decreasing—during puberty, in the days before menstruation, after giving birth to a baby, or before and after the onset of menopause. Not coincidentally, depression is also associated with sleep problems. People with sleep problems are much more likely to be depressed.

As complex as an orchestral piece, the menstrual cycle requires the proper sequencing of hormones and activities in at least four different tissues of the body: the hypothalamus and pituitary gland, which are in the central nervous system; the ovaries; and the uterus. The cycle is made up of three distinct phases:

1. *Follicular phase.* One of the dormant eggs in the follicle in the ovary develops, and at the same time the lining of the uterus begins to prepare itself to nourish a fertilized egg.
2. *Ovulation.* On day fourteen (in most women), midway through the monthly cycle, the egg is released and makes its way into the fallopian tube.
3. *Luteal phase.* The uterine lining thickens in preparation for possible fertilization. If fertilization does not occur, the lining of the uterus is shed. This causes the menstrual bleeding. The cycle then repeats.

Every woman is intimately familiar with the rhythms of her menstrual cycle, but not every woman realizes how the three phases can affect the quality and quantity of her sleep. Simply put, the level of hormones fluctuates intensely, swinging between low and high levels each month in women of childbearing age, and affecting many tissues of the body, including the nervous system, which controls sleep. Disruptions in sleep can occur during the regular menstrual cycles, and more serious sleep problems can occur in three conditions linked to hormonal changes: premenstrual syndrome (PMS), premenstrual dysphoric disorder (PMDD), and polycystic ovarian syndrome (PCOS).

Sleep Patterns of Women with Normal Menstrual Cycles

For most women who have regular periods, the menstrual cycle is not associated with sleep complaints, although research studies have found that there are subtle changes in the amount of sleep women get and in the levels of daytime sleepiness they experience. Research published in 2016 reported that the most frequent sleep disruption occurs when hormone levels change the most quickly: during ovulation and immediately before menstruation. At these times women may experience a few nights of sleeplessness. Some women, however, do not notice a change in their sleeping patterns, or might feel only slightly sleepier. Women with irregular cycles experience greater sleep disruption.

Women typically sleep the most during the early follicular phase. In the days before ovulation, the estrogen levels increase and the amount of rapid eye movement sleep also increases slightly. Women sleep the least during ovulation. This is probably because of the effect on the brain of high levels of hormones that cause ovulation.

In the third phase of the cycle, there is an increase in progesterone, which causes the body temperature to rise. The number of brief awakenings is highest in the few nights right before bleeding starts, when levels of progesterone and estrogen are both dropping. Women might find themselves sleeping fitfully. In the late luteal phase, many women report that it takes them longer to fall asleep, they sleep less, and the quality of the sleep is poor compared with that at the beginning of the cycle.

Some otherwise normal women may develop very severe incapacitating sleepiness during menstruation. Birth control pills can be effective in preventing the sleepiness.

Some women have painful cramps before and during menstruation, which can cause wakefulness. Many women find it hard to get to sleep during this stage, and after they fall asleep they tend to have less REM sleep and a slightly elevated body temperature. Such women are likely to be sleepier than normal.

Women who take birth control pills, which control the menstrual cycle, may have different sleep patterns from those who use other types of contraception. The main effect of the pills is to prevent ovulation, so women who use birth control pills may not experience the mild effects on sleep related to ovulation. These women may still have symptoms related to menstruation, but some women who have severe sleep difficulties during menstruation might find that the symptoms were less severe than they were before they began using the pill.

Sleep and Other Problems Related to PMS or PMDD

Women with PMS exhibit a variety of symptoms (trouble sleeping, irritability, mood changes, bloating) before menstruation. According to a 2012 study, about 76 percent of women with PMS experience sleep difficulties.

Most women who have PMS experience the symptoms a few days before the start of menstrual bleeding (the late luteal phase). Usually the symptoms end when the bleeding starts or within two to three days afterward.

Scientists have been unable to pinpoint a single mechanism that causes PMS because many hormonal and chemical changes occur before menstruation, and each woman with PMS probably has a more or less unique combination of symptoms caused by a unique combination of chemical changes. Because the symptoms vary so greatly, doctors should consider other potential disorders that have similar symptoms when diagnosing PMS. Symptoms such as sleeplessness, hot flashes, and a rapid heart rate might be found in hyperthyroidism (excess thyroid production) or menopausal transition. Tiredness could be caused by hypothyroidism (inadequate thyroid production). Some experts believe that in a small number of women, abnormal thyroid function may lead to symptoms of PMS. And in some women, the sleep problems and mood swings associated with PMS might actually be symptoms of depression.

We can divide the symptoms of PMS into two general categories: those that affect the nervous system and those that affect other parts of the body. The nervous system symptoms include problems sleeping (which can be severe), mood swings, irritability, anger, headaches, memory loss, and tremors. The symptoms involving other parts of the body include breast swelling, fluid retention, muscle aches, nausea, and vomiting. Most women with PMS have only a few of the symptoms, however, though many have trouble sleeping and may experience daytime sleepiness.

For a diagnosis of PMS, the symptoms should be present over several consecutive menstrual cycles and be severe enough to interfere with the woman's mental state and activities of daily living. Because the range of symptoms affecting almost all the organ systems of the body is staggering, a diagnosis of PMS is still a difficult one to make. As yet we have no standard test that can confirm such a diagnosis. Women may go to various doctors, sometimes for years, before their PMS is diagnosed.

Because women with PMS exhibit so many differing symptoms, the treatment is not specific to the syndrome (whose cause is still unknown) but is based on the symptoms themselves, with the expectation that the symptoms will disappear after menstruation begins.

Three approaches to the type of medication to help PMS and its effect on sleep are common:

- Relief of specific symptoms such as pain
- Change of hormone levels
- Prevention of the mood disorders before they occur

If a woman has been experiencing pain (tender breasts, severe cramps), the doctor might suggest an over-the-counter pain medication that has anti-prostaglandin properties (also called nonsteroidal anti-inflammatory medications or NSAIDs). In the United States, examples of NSAIDS available over the counter include those containing ibuprofen (Advil, Motrin, Nuprin, and Midol 200) and those containing naproxen (Aleve). Generic versions of these medications are also available. Check with a pharmacist before choosing a medication. For bloating and water retention, a doctor may prescribe a mild diuretic. These medications are not taken daily, only when symptoms are severe.

The medications that change mood and hormone levels are powerful and have potentially severe side effects, and women should take them only in consultation with their doctor. Antidepressants approved by the U.S. Food and Drug Administration (FDA) for the treatment of severe PMS and PMDD include Prozac (fluoxetine), Zoloft (sertraline), and Celexa (citalopram). The FDA has approved a contraceptive medication called Yaz as well to treat the symptoms of PMS. I don't recommend any of these drugs if the main or only symptom being treated is sleeplessness. If the sleeplessness is caused by PMS, it will improve in a few days after bleeding begins.

The effect of long-term use of these drugs on PMS is not known, so the patient should discuss the pros and cons of the drugs with her doctor. Remember, the symptoms generally disappear once menstruation begins, so the best approach might be to take nothing. This is especially true if the woman is planning on getting pregnant in the near future. If the woman becomes pregnant while taking these medications, she should contact her doctor at once.

Medications are not the only way to counteract sleeplessness caused by PMS. Women who notice that it is taking them much longer than usual to fall asleep or who are waking up frequently at night should reduce caffeine intake as a first step. (At the very least, they should stop drinking coffee or tea after lunch.) Similarly, although many people believe that alcohol will help them fall asleep, it can also cause them to wake up later on in the night, disrupting their sleep. Women with PMS should therefore avoid alcohol at night. Women who have repeatedly experienced disrupted sleep during their menstrual cycle often expect to have a bad night's sleep before menstruation. The expectation of a bad night causes stress and can itself lead to a bad night. Women may benefit from learning relaxation methods that reduce the stress caused by the expectation of a bad night. If the sleep problem is very severe and does not respond to the treatments for PMS, the sleep problem may not be related to PMS. In that case, other types of problems should be considered. If the sleep difficulty remains severe, the woman should seek help from a gynecologist.

If mood and nervous system premenstrual symptoms are very severe, they might indicate a more serious problem, premenstrual dysphoric disorder (PMDD). If, along with the symptoms of PMS, a woman experiences symptoms of depression (hopelessness, severe sadness, or thoughts of suicide; see Chapter 16) or anxiety, wide mood swings, severe uncontrollable anger or irritability, and marked problems with sleeping, she may be suffering from PMDD. A woman with PMDD might experience severe insomnia and have extreme difficulty falling asleep and staying asleep. She might awaken very early in the morning and not be able to fall asleep again. Some patients with bipolar disease are misdiagnosed as having PMDD. Doctors whose patients experience severe problems that worsen before menstruation should evaluate these symptoms carefully when making their diagnoses. Up to 75 percent of women with PMDD notice an improvement in their symptoms with antidepressant treatments. Research from Japan reported in 2016 showed that increasing fish consumption improves performances in athletes with PMDD!

Sleep and Other Problems Related to PCOS

Most tissues that produce one hormone are capable of producing other chemically related hormones. In most women with polycystic ovarian syndrome (PCOS), the ovaries produce too much of the male sexual hormones (androgens). Research reported in 2015 suggests that women can have one of four variations of PCOS. In one type, there is excess male hormone production with abnormal ovulation and ovarian cysts. In the second type, excess male hormone production also occurs, accompanied by abnormal ovulation *but without ovarian cysts*. In the third type, excess male hormone production is accompanied by ovarian cysts but *normal ovulation*. In the fourth type, ovarian cysts and abnormal ovulation occur but *without increased male hormone production*. It is likely that in women with the second type (without ovarian cysts) fat cells might be producing the excess male hormone.

When there are high levels of male hormones in women, this may lead to low levels of follicle-stimulating hormone. As a result, the eggs in the follicles might not develop. The follicles swell and form collections of fluid called cysts, and many follicles with undeveloped eggs can form these cysts, hence the name "polycystic." The ovaries sometimes increase in size dramatically, becoming as large as a baseball, or even larger. These abnormal hormone levels cause two sets of problems: as shown in the twenty-nine-year-old patient, women may develop excess hair and other features normally found in males, or they might experience problems with their reproductive system. Such severe symptoms might seem unusual, but PCOS is in fact a common disorder, found in about 5 to 20 percent of premenopausal women. In about a quarter of teenage girls who don't menstruate, PCOS is the probable cause.

The most common symptoms of PCOS are male hair distribution, overweight, and problems with the menstrual cycle or difficulty in becoming pregnant. The problem may first become apparent when a woman is being investigated for infertility. Women with PCOS, for example, may have facial hair or develop acne well past the teenage years, in their twenties and thirties, and might even develop baldness. Infrequent menstrual cycles and even complete cessation of menstruation are common in PCOS patients. Women with PCOS also develop a resistance to the effect of the hormone insulin, which usually lowers blood sugar. This can lead to diabetes in about 10 percent of women with this condition and an increased risk of cardiovascular disease. These women also have abnormal blood lipids, which increases their risk for heart disease; because of their excess weight and the male distribution of the extra weight, they are much more likely to develop obstructive sleep apnea. (It is not just the weight that leads to the apnea, but the location of the fat tissue. Women with PCOS have a male fat distribution—that is, the waist increases more than the hip size. PCOS women have a larger waist-hip ratio than the other women and much higher levels of testosterone.)

If a woman is obese and has acne and facial and body hair in a distribution normally found in men, PCOS should always be suspected. Abnormal or absent periods and infertility are common. Diabetes, hypertension, and sleep apnea may also be present. In one study reported in 2014, 66 percent of women with PCOS were shown to have a sleep-breathing disorder. These patients snore, stop breathing during sleep, and experience daytime sleepiness. Women with PCOS who suffer from sleep apnea are also much more likely to have metabolic problems and nonalcoholic fatty liver disease.

Women who suspect they have this condition should see a doctor, especially if they want to become pregnant. Losing weight can be very effective in helping to manage the hormonal changes, and may also help alleviate the sleep-breathing problem. Metformin, a medication often prescribed for diabetics, increases the response to the insulin produced by the body; this can improve the symptoms of PCOS and may normalize the menstrual cycle. For some women, metformin can also lead to weight loss, which leads to a decrease in male hormone production and an improvement in the body's ability to respond to insulin, important in controlling diabetes. Sometimes even a relatively small weight loss can lead to dramatic improvement in the chance for a successful pregnancy or normal menstruation; it may also help relieve sleep apnea. If the patient is unable to lose weight, an effective treatment to relieve sleep apnea is continuous positive airway pressure (CPAP) treatment, which I discuss in Chapter 12. Patients wear a mask attached by a hose to a device that generates pressure and keeps the breathing passage open. Recent research suggests that the apnea may itself reduce the effectiveness of the hormone insulin. Reduced insulin levels or decreased insulin effects play an important role in diabetes, and treatment of the apnea in PCOS may improve diabetes in these patients.

Back to the the Sleepy Woman with Irregular Periods

My twenty-nine-year-old patient's irregular periods were caused by a disease that made her ovaries produce too much male hormone (hence the irregular periods) and resulted in a disorder that is more common in men than women: sleep apnea. An overnight sleep test showed that she stopped breathing repeatedly, about once a minute, while she slept. When she was asleep the muscles in her throat relaxed and the upper breathing passage became blocked, which caused her breathing to stop. Each time this happened, the level of her blood oxygen dropped to a dangerously low level, and her brain would wake her up in order to open up the breathing passage and start her breathing again. This happened hundreds of times during the night, and was the cause of her snoring, her severe sleepiness, and her headaches.

Cysts in the patient's ovaries had produced too much male hormone. She had the symptoms of PCOS. The male hormone also caused her to have abnormal periods, to be overweight, and to have a male distribution of body hair. To relieve the immediate problem of daytime sleepiness and headaches and improve her quality of life, she was started on CPAP, and as a result her morning headaches stopped within days. She is now seeing a gynecologist to manage her menstrual problems and trying to lose weight, and she feels great. The sex hormones affect women's bodies in profound ways, and for this patient, abnormal sex hormones had caused a sleep disorder that had jeopardized her life.

Most women (about two-thirds) experience some form of sleep disturbance linked to menstruation. Although medical science has learned to better understand and help women with these problems, women should be aware that menstruation causes sleep problems and that PMS can make sleep problems worse. Nowadays, however, these sleep problems can be treated once they have been identified.

4

Pregnancy and Postpartum

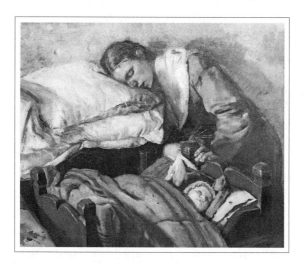

THE MYSTERY. Most women have sleep problems during pregnancy and after the birth of the baby. Sleeplessness can be severe, often caused by disorders such as restless legs syndrome, preeclampsia, and sleep apnea.

The Case of the Sleepy New Mother

Three months after giving birth to her first baby, a twenty-nine-year-old woman came to see me at the sleep clinic in a state of total exhaustion. Although most new mothers can expect to be fatigued by the demands of caring for a newborn, this woman's lack of restful sleep was affecting the quality of her life and the care she gave her baby— she could not stay awake long enough during the

day to care for him properly. She fell asleep while watching television or reading, during conversations, and even sometimes when driving her car, in spite
of drinking six to eight cups of coffee a day.

More disturbing was the fact that her sleep problems had started during
pregnancy, and because they went untreated she and her doctor had unknowingly put her life and the life of her child at risk. I was not surprised to find
out that her full-term baby was underweight at birth—only five pounds, ten
ounces—and that she had had two previous miscarriages. As I took her history,
more details emerged. The patient had been thirty or forty pounds overweight
before her pregnancy and had experienced daytime sleepiness for about ten
years, but the sleepiness had become much more severe once her pregnancy
began. When she was six months pregnant, she spoke to her doctor about this
problem. She explained that she had become uncontrollably sleepy, often
falling asleep even when she did not want to. Her snoring had also become
much worse since her pregnancy began, and her husband had noticed for at
least five years that she stopped breathing when she was asleep. Her doctor
told her that it was possible that she had sleep apnea. But the doctor also
advised her not to get any tests or treatment for her sleep disorder during pregnancy, explaining that it would be much safer to wait until the baby was born.
Like many doctors, hers did not understand the serious health consequences
of women's sleep disorders.

Sleep Disorders During and After Pregnancy

Unfortunately for mothers-to-be, sleep problems are a normal part of pregnancy. A study from 2014 suggests that chronic sleep loss in pregnant women
can result in cardiovascular and kidney problems in their offspring. The 1998
National Sleep Foundation poll on Women and Sleep found that almost 80
percent of women report more disturbed sleep during pregnancy than at other
times. Most of these women mentioned their frequent need to urinate as the
main reason. Other reasons were related to various symptoms of pregnancy:
tiredness, pelvic pressure, insomnia, lower back pain, restlessness, leg cramps,
and nightmares. Symptoms of chronic sleep disorders, such as restless legs
syndrome or sleep apnea, can also appear at this time or can worsen. These
symptoms vary within the three trimesters of pregnancy.

FIRST TRIMESTER

When they first become pregnant some women feel wonderful. Others feel awful. Some women's sleep remains normal, while other women sleep poorly from the beginning of their pregnancy and find themselves becoming more tired during the day. As one twenty-nine-year-old first-time mother reported, "I found many short rests during the day and night was all I could do to combat my sleepiness. I don't think there was one night that I didn't get up about six or seven times to use the washroom or to get water or simply could not sleep while I was pregnant. I would sleep an hour, then awaken every hour during my whole pregnancy." This difficulty sleeping is probably due to the effect on the brain of rising progesterone levels. Morning sickness, which is common in the first twelve weeks of pregnancy, can also cause women to awaken with severe nausea. The nausea and vomiting can be extreme and become a serious health risk, as was the case for Kate Middleton, the duchess of Cambridge. This condition is now called "nausea and vomiting of pregnancy." The only medication approved by the FDA to treat it is a delayed-release combination of doxylamine succinate and pyridoxine hydrochloride (Diclegis).

SECOND TRIMESTER

During the second trimester many women experience fatigue or tiredness due to the demands on the body, the carrying of extra weight, and sleep problems related to the enlarging uterus. Some may find that though they are tired and want to spend more time in bed, they have trouble falling asleep and spend much of the night restlessly tossing and turning, trying to find a comfortable position. A more extreme version of this restlessness is a medically diagnosable sleep disorder called restless legs syndrome, an uncontrollable urge to move because of unpleasant tingling sensations in the legs. Some women develop other problems that keep them awake at night such as cramps in their calves or back pains.

In the second trimester, women may also start to experience heartburn at night, which may continue until the end of pregnancy. The heartburn is caused by acid from the stomach going back up into the esophagus, the tube that carries food from the throat to the stomach. One contributor to acid reflux is the extra pressure on the stomach caused by the enlarging uterus. There are also muscles at the bottom of the esophagus that normally keep acid from backing up from the stomach. These muscles might not work as well during pregnancy. Having a snack, especially spicy foods, two or three hours before bedtime can bring on or exacerbate this condition.

THIRD TRIMESTER

During the last trimester of pregnancy, a wide range of problems can disrupt sleep. Some women develop nasal congestion, which may cause them to snore or develop the symptoms of sleep apnea. Some women have a marked worsening of restless legs syndrome. Others become breathless if they lie in certain positions or experience severe back pains that interfere with their sleep. Closer to the birth, breathing may also become more difficult because the uterus enlarges to the point where it pushes up on the diaphragm, the major breathing muscle. As pregnancy continues, the woman's sleep becomes more and more difficult because the discomfort and sensations caused by the baby's movements can lead to general overall restlessness. It is not uncommon for an expectant mother to stay awake all night toward the end of her pregnancy. I have seen some women who are completely unable to sleep twenty-four hours before going into labor. (Some also start to demonstrate nesting behavior: one woman used a sewing machine for the first time in her life on the night before her daughter was born and sewed curtains for the new baby's room. She never used the sewing machine again!) Some scientists have suggested that the poor quality of sleep women may experience during pregnancy is the body's way of preparing them for spending a great deal of time taking care of the newborn at night.

MULTIPLE-BIRTH PREGNANCIES

The sleep problems pregnant women encounter are magnified for women who carry more than one baby. Giving birth to several babies has become more common owing to advances in fertility treatments. And because the uterus can enlarge dramatically in women carrying several babies, the discomfort can be greater. Most women who are pregnant with more than one baby can only sleep on their sides. They cannot lie on their stomachs, and sleeping on their backs often results in breathing difficulties. The nutritional demands of the developing babies make the mother more likely to develop an iron or vitamin deficiency. Mothers carrying several babies often need to go to the hospital or be placed on bed rest before their due date to prevent giving birth prematurely.

SLEEP DISORDERS FOLLOWING
THE BIRTH OF THE BABY

When the mother gives birth she experiences a dramatic drop in her progesterone level, at the same time that other hormones are starting to kick in so that she will be able to breastfeed. After the baby is born, if there are no complications for the mother or the baby, the mother's sleep can return to normal fairly quickly, although "normal" is a relative term while she's doing frequent night feedings. During delivery, excessive blood loss can occur. Additionally, some women develop an iron deficiency during pregnancy because the developing baby saps iron from the mother. The iron deficiency combined with the blood loss can lead to anemia (low red blood cell count). This can result in severe daytime fatigue, especially if the new mother is also suffering from sleep deprivation because she is awakened for nocturnal feedings or by the baby's crying. If the baby was delivered by caesarean section, the mother may also experience substantial pain and discomfort, which can contribute to poor sleep.

Dealing with Sleep Problems in Pregnancy

INSOMNIA

In addition to the two sleep disorders that are often seen in pregnant women—sleep apnea and restless legs syndrome—the normal discomfort, pain, and varied sensations associated with feeling the baby develop inside her can cause the mother to experience insomnia. This can be dangerous, for lack of sleep has been associated with a higher rate of miscarriage. It is important for expectant mothers to get the sleep they need during pregnancy. In Chapter 10, I offer suggestions for sufferers from insomnia, but some need to be modified for pregnant women.

First of all, I must caution pregnant women seeking relief against using some remedies that they might have used to combat insomnia before pregnancy. Sleeping pills, alcohol, or over-the-counter medications and herbals are not recommended to help pregnant women sleep through the night. Scientists do not know what long-term risks to the baby these medications or supplements might entail. But there are other options, involving behavior modification, open to pregnant women suffering from insomnia that can offer them some much-needed relief.

Short naps can be extremely helpful. The emphasis should be on short—

too long a nap might interfere with nighttime sleep. A good time to nap is in the early afternoon, and the naps should be no longer than twenty to forty minutes.

Women who get heartburn at night should avoid spicy food, acidic fruit juices, and alcohol. They should not eat big meals and probably not eat anything in the two to three hours before they go to bed. Even naps should be put off until at least a half hour after eating. If the heartburn is severe, women may want to change their position during sleep, leaning against several pillows or even resting in a recliner. Women whose heartburn seriously interferes with their sleep should consult a doctor, who might prescribe an antacid such as Tums, which will have the added benefit of providing calcium. A glass of skim milk could offer temporary relief, but pregnant women should avoid whole milk, which can cause more acid to form. Expectant mothers should be careful not to drink too much milk to relieve heartburn as the calories can lead them to put on extra weight, which can in turn lead to sleep apnea.

Perhaps the most effective way for pregnant women to fight insomnia is to experiment with sleeping positions until they find the most comfortable one. Most women have to train themselves to sleep in a different position from those they used before pregnancy. For instance, women who tended to sleep on their stomachs can no longer do this and women who slept on their backs might now find it too difficult to breathe. Sleeping flat on the back might also result in the uterus pressing on the main artery of the woman's body, the aorta, which could reduce blood flow to the baby. For many women, sleeping on the side becomes the most comfortable position; it will also increase blood flow to the baby by relieving pressure on the aorta. Women who are not used to sleeping on their sides might find that putting a pillow between their knees can improve their comfort level.

Unfortunately, there is no absolute remedy for pregnancy-related sleeplessness except giving birth.

RESTLESS LEGS SYNDROME

Restless legs syndrome (RLS) frequently occurs during pregnancy. (See Chapter 11 for a fuller discussion of RLS.) Many women with RLS recall that their first episode of the disorder occurring during a pregnancy. Women with RLS feel an irresistible urge to move their legs while they are trying to sleep. Moving or walking relieves this urge. In some patients, RLS is inherited. Women who have inherited RLS are much more likely to have increased symptoms of restlessness and a crawly sensation under their skin during pregnancy.

A study done at the University of California, published in 2001, focused on the likelihood of women developing RLS during pregnancy. None of the women studied had the syndrome before they became pregnant, but by the end of their pregnancies, 23 percent of them were afflicted. They also experienced insomnia and a depressed mood. Those who developed RLS were found to have iron deficiencies and/or folic acid deficiencies before becoming pregnant, both known causes of RLS. Women with an iron deficiency should have it treated. So common sense—as well as research reported in 2015— mandates that women who develop restless legs syndrome during pregnancy should have their iron status checked by a doctor, who will probably prescribe a multivitamin preparation containing folic acid. Pregnant women need to keep their folic acid levels up for another reason as well: studies have shown that mothers who take folic acid are less likely to have children born with a neurological malformation.

RLS can be miserable for pregnant women. They may develop restlessness in their legs, not only at night but during the day as well. A detailed review published in 2015 summarized treatment options and concluded that *not one drug treatment for RLS was both safe and effective* during pregnancy or while the mother was breastfeeding. The treatments that could be considered were mild to moderate exercise (most women learn that walking often helps), massage, pneumatic compression devices, iron therapy, and yoga, as well as treatment of sleep apnea if that were also present. One pneumatic compression device consists of a wrap around the lower leg that is inflated for five seconds every minute. This is used for an hour in the evening. Similar devices that are used to prevent blood clots in the legs might also be effective. The FDA has approved a device (Relaxis) that applies vibrations to the legs.

In most women, restless legs syndrome goes away with childbirth, although for some women the symptoms brought on by pregnancy can go on indefinitely. These women may benefit from seeing their doctor.

SNORING

About a third of pregnant women snore. Since women are less prone to snoring than men, often the first time a woman snores is when she is pregnant. A 2015 study reported that pregnant women who snore and otherwise score high on a sleep apnea questionnaire are more likely to develop high blood pressure and to need a cesarean section. Some research studies have shown a frightening connection between snoring and a potentially dangerous medical condition called preeclampsia, a form of high blood pressure that can occur during pregnancy. Preeclampsia causes damage to the kidneys, which results in a great deal of protein being lost in the urine. Studies have found that pregnant women who snore are twice as likely to develop preeclampsia than those who do not snore, and women with preeclampsia are twice as likely to have smaller babies than those without it. Affecting about 7 percent of pregnant women, preeclampsia usually begins after twenty weeks into the pregnancy. Blood pressure goes up, causing damage to the kidneys. Women may experience few obvious symptoms, as high blood pressure and kidney problems can be difficult to detect without testing, though early in the pregnancy sufferers might have headaches or severe swelling of the ankles, or might have swelling of the face. The most obvious indicator is often snoring. (Some women who snore during pregnancy might have sleep apnea, which is discussed below.)

Thus, if a woman begins to snore while she is pregnant, she should have her blood pressure and urine checked, especially if she also suffers from headaches and swollen ankles. About one in twenty women with preeclampsia develops seizures as well as very severe high blood pressure and other problems. This even more serious condition is called eclampsia. Both conditions usually resolve soon after childbirth. Women with preeclampsia or eclampsia require immediate medical care, and some might benefit from a sleep study.

The dangers these conditions present highlight the importance of talking with a doctor about sleep problems—even seemingly insignificant ones such as snoring.

SLEEP APNEA

Some women experience sleep apnea before they become pregnant, and the condition worsens during pregnancy. In other women, the apnea develops during pregnancy. Screening for apnea by questionnaire can be useful in the second and third trimesters. The symptoms are similar to those of sleep apnea in women who are not pregnant: snoring, pauses in breathing during sleep, and severe daytime sleepiness. Some of the medical issues that occur during pregnancy have now been linked to sleep apnea in the mother. These include high blood pressure and gestational diabetes.

Research reported in 2010 and 2013 showed that when a pregnant woman has symptoms of sleep apnea she is more likely to have pregnancy-induced hypertension, preeclampsia, and gestational diabetes, or to require a cesarean delivery. Research reported in 2016 showed that mothers with sleep apnea are at greater risk of giving birth prematurely. The baby is liable to be less vigorous or to need to spend time in a Neonatal Intensive Care Unit. Some babies of mothers with sleep apnea might be large for their gestational age, and thus more prone to hypoglycemia, whereas others might be smaller than average at birth and are more likely to develop a metabolic disorder, or even cardiovascular disease, in adulthood.

It has also been suggested that women with sleep apnea are more likely to have miscarriages. The danger of sleep apnea is that the blood oxygen level in the mother can drop to very low levels. Because the baby is dependent on the mother for oxygen, a pregnant woman with sleep apnea must be treated as soon as possible. The usual treatment is continuous positive airway pressure (CPAP), a mask worn over the nose connected by a hose to a machine that increases pressure in the breathing passage. The system can be used by people who sleep on their sides, as many pregnant women do.

In addition to the risks a pregnant woman's sleep apnea presents to her fetus are the problems it presents for her not only during pregnancy but after the baby is born, when the new mother will have many responsibilities. Treatment of her sleep apnea can help ensure that she gets a good night's sleep and be wide awake and alert during the day to cope with the demands of her baby. Dealing with the needs of a new baby is not easy for any woman, but it is extremely difficult for a woman with untreated sleep apnea. The first pregnant woman I saw who had sleep apnea had to be started on emergency treatment immediately after delivering the baby because there was no one else to care for her child.

Postpartum Sleep Problems

SLEEP DEPRIVATION

The sleep difficulties seen during pregnancy are often followed by sleep deprivation after the baby is born. About 20 percent of immediately postpartum mothers may have sleep apnea. The new obligations involved in caring for the baby eat away at the mother's time. For breastfeeding mothers, one of these is waking up at night to feed the baby. Often a newborn baby sleeps in the parents' bedroom or bed, and can disturb new mothers who are trying to sleep. Mothers who can get outside help should do so. They should also make a point of napping when the baby naps. Finally, new mothers should be careful not to drive when sleepy.

MOOD CHANGES AND DEPRESSION

Although new mothers are usually euphoric after giving birth, some women have changes in mood that can range from feeling temporarily blue to having an episode of full-blown clinical depression. Sleeplessness is a common symptom of depression. It is believed that these abnormal moods are caused by reduction in the levels of the hormone progesterone. Women who have severe depression after childbirth often have had previous depressive episodes, although those episodes might not have been diagnosed. Postpartum depression can have devastating results (for example, depressed mothers might commit suicide or harm the newborn), and new mothers who have depressive symptoms should receive immediate medical attention. I discuss these symptoms fully in Chapter 16.

Back to the Sleepy New Mother

We gave the patient a sleep test to measure how many times she stopped breathing while she slept. The test revealed that she had severe obstructive sleep apnea syndrome, which caused her to stop breathing 136 times an hour, and almost every time her breathing stopped, her brain awakened for a few seconds. Stopping breathing more than 5 times an hour is considered significant, so this new mother's sleep apnea was extremely severe. When breathing stops, the amount of oxygen in the blood decreases. Correspondingly, her blood oxygen level dropped to dangerous levels; she spent about 11 percent of her sleep time with her blood oxygen level below 80 percent. During these episodes, 20 percent of her blood was not carrying oxygen, which put great stress on her cardiovascular system. (These oxygen levels are typical in people sleeping at fifteen-thousand-foot altitudes but abnormal for those at sea level.) The sleep apnea was putting her life and the life of her child at risk and was the probable cause of the baby's low birth weight.

After the first sleep test, which confirmed her severe sleep apnea, the sleepy new mother was tested on a CPAP mask. The result was a dramatic improvement—her apnea episodes disappeared. Her sleep became completely normal and her blood-oxygen levels never again decreased to their previously dangerous lows. Two months later, after using the CPAP at home every night, she told me that she felt great because she was no longer sleepy. "I can't sleep without [the CPAP]," she said. "Everyone says I'm better." With the assurance of a good night's rest, she felt better able to contend with her daily responsibilities.

Women need to become more aware of the impact of sleep disorders on their health. Although pregnancy is often a time of great joy and excitement for a woman, it can also seriously affect her sleep. The majority of women have problems with sleep during pregnancy, ranging from mild to severe. It is important that women and their doctors know about sleep problems that affect women during normal, multiple-birth, and high-risk pregnancies.

Most sleep problems that occur during pregnancy improve after the birth of the baby. The end of a woman's reproductive years, however, do not signal the end of potential sleep problems. After the onset of menopause, women often find themselves experiencing sleep problems. In the reproductive years, fluctuating sex hormone levels are often the cause of sleep disturbances; in the menopausal years, it is the lack of sex hormones that cause the problems.

5

When Sex Hormone Levels Decrease

MENOPAUSE AND ANDROPAUSE

THE MYSTERY. Changing hormone levels affect women's sleep not only during menstruation and pregnancy but during menopause, when the levels of sex hormones drop. This causes a number of symptoms that affect sleep, including hot flashes and night sweats, as well as symptoms of sleep apnea. Changing hormone levels can also affect men being treated with drugs to reduce levels of testosterone for certain cancers and women being treated for breast cancer, and can lead to similar symptoms.

~~~~~~~~~~~~~~~

The Case of the Insomniac with Night Sweats

Sitting in front of me was a fidgety, thin, anxious fifty-one-year-old. Her family doctor had referred her to me because she had been complaining of experiencing a great deal of difficulty in both falling asleep and staying asleep. Most nights, after tossing and turning in bed trying to get comfortable, she would drop off but then, frustratingly, would not be able to sleep through the night. Instead, she would find herself waking up several times, often covered in sweat, with the back of her head and her pillow soaked in perspiration. Sometimes she awakened with her heart pounding, which frightened her. As a result of her interrupted nighttime sleep, she would feel exhausted during the day. Her problems sleeping were affecting every aspect of her life.

She could not pinpoint a particular cause for her sleeplessness. She had no special problems at home or at work, nor could she attribute her sleeplessness to problems with her mood.

Her family doctor had suggested sleeping pills, but she did not like taking pills and preferred to find an alternative way of treating her sleep problems.

She was becoming more and more worried about her sleeplessness, which was causing her to lose weight, and her anxiety further exacerbated her difficulties in falling asleep. Only a few conditions commonly cause this constellation of symptoms.

Menopause

Menopause, the period following the reproductive phase of a woman's life when her menstrual cycles have ceased and her body no longer produces estrogen, is a transition experienced by all women. Menopause does not usually come on abruptly; rather, menstrual cycles become irregular or more time elapses between periods. The amount of bleeding with each period may vary. Most doctors agree that menopause is established if periods have been absent for a year. Menopause is not a disease; it is a normal physiological state.

The age at which menopause occurs varies. In some women, it may start when they are in their early forties, while for other women it might not begin until they are over fifty. In North America, most women experience the onset of menopause sometime between the ages of forty-eight and fifty-five, with the average age being about fifty-one. Women might start menopause early because their ovaries were surgically removed for medical reasons. Women being treated for breast cancer are much more likely than others to have menopausal symptoms; the symptoms are particularly common in women who have been treated with medications such as tamoxifen, an anti-estrogen that counteracts the effects of the female hormone estrogen on breast cancer cells, or an aromatase inhibitor, which blocks the production of estrogen (there are several that have been approved in the United States by the Food and Drug Administration).

Just as dramatic changes occur in organ systems for women during adolescence when the ovaries start to produce estrogen, the abrupt reduction of estrogen production during perimenopause and menopause also results in a wide variety of effects. The five most disturbing menopausal symptoms are hot flashes, vaginal dryness, night sweats, disrupted sleep, and weight gain. All involve sleep—even the weight gain could lead to sleep apnea. (These symptoms can also affect the sleep and quality of life of the bed partner.) When menopause occurs abruptly— for example, after surgical removal of the ovaries—symptoms can be quite severe. Some women experience relatively few symptoms with menopause.

HOT FLASHES

During menopause, the way a woman's body regulates its temperature can change, often to her extreme discomfort. The hot flash is one of the most unpleasant symptoms of menopause, and it is experienced by between 80 and 90 percent of perimenopausal and menopausal women. A study published in 2003 showed that women who smoke cigarettes or are obese are twice as likely as nonsmokers of normal weight to have severe hot flashes. When a woman experiences a hot flash, she feels as though her body temperature is increasing. Indeed, it does increase by a small amount, which fools the hypothalamus, the part of the brain that regulates body temperature. Many scientists believe that reduced estrogen levels, especially if they are reduced rapidly, and release of certain hormones from the pituitary gland cause the hypothalamus to respond as if the body is overheated. This in turn activates the mechanisms the body uses to rid itself of excess heat, redirecting blood flow to the skin and causing sweating.

The main mechanism the body uses to get rid of extra heat is a process called vasodilatation (enlargement of the blood vessels). Blood vessels enlarge, and the blood flow to the skin is increased. So even though the woman feels hot, her body is actually losing heat. It is this increase in blood flow that results in the sudden flushing sensation known as a hot flash. Many doctors use the term *vasomotor symptoms* to describe the features of the hot flash. The episode usually begins with the perception of feeling hot, followed by flushing of the face, which may then spread to elsewhere on the body. Some women say that the flush begins at their chest and then moves up. The average episode is roughly three minutes long and causes extreme discomfort.

Usually women experience hot flashes for 1 to 7 years, but some women experience them for more than 10 years. A 2015 study reported that the average duration varied by race and ethnicity: African American women averaged 10.1 years, Hispanic women 8.9 years, non-Hispanic Caucasian women 6.5 years, Chinese women 5.4 years, and Japanese women 4.8 years. Many women who experience hot flashes have more than ten episodes per day, which can disrupt their home and work lives. Other women have episodes as infrequently as once a month.

When the hot flashes come at night, the woman may experience night sweats that adversely affect her sleep. (They might also interfere with her bed partner's sleep; as the husband of a perimenopausal woman with hot flashes noted, "At first, I did not realize my wife had a problem at all. I thought that I had a problem. I was waking up feeling really cold in the middle of the night and found, when I checked, that the thermostat was set much lower than normal. After a few nights of this I discovered that my wife had been setting the temperature much lower.") Recent research has shown that hot flashes do not occur during REM or dreaming sleep; during REM sleep the body no longer controls its temperature, and abnormal temperature regulation is what leads to the hot flashes.

At the end of the hot flash, women often break out into a sweat over the part of the body that had been involved in the flush. When these episodes occur while a woman is asleep, as they frequently do, the profuse sweating may bother her so much that she cannot get comfortable; she might even have to change the bedclothes if the night sweats are too severe. To make matters worse, at the end of the hot flash, the hypothalamus registers that it has cooled the body too much, and so it turns on the mechanisms to increase the body temperature, which can make the woman feel cold and clammy.

OTHER PHYSICAL EFFECTS OF MENOPAUSE

Hot flashes are perhaps the best known and most familiar symptom of menopause, but the reduction of estrogen and progesterone can have other effects on a menopausal woman's body as well. Estrogen is crucially involved in the reproductive system, and when the production of estrogen decreases the walls of the vagina become thinner, while the production of lubricating fluid is also decreased. The resulting dryness in the vagina may make sexual intercourse painful. Another effect of menopause for many women is a dramatic metabolic change. Often this manifests as weight gain, which among other dangers may increase the likelihood of or put a woman at risk for developing sleep apnea.

Besides estrogen, another important sex hormone that decreases with menopause is progesterone, a hormone that is believed to help protect against the development of sleep apnea. The combination of increased weight and decreased progesterone dramatically increases the risk of a woman's developing sleep apnea. Finally, the loss of estrogen during menopause puts women at increased risk for cardiovascular disease, cancer (uterine, breast, or ovarian), bone fractures, and other conditions. The cancer risk seems to increase if menopause occurs after age fifty-five.

One often overlooked issue is the impact of all the physiological changes that occur during menopause in conjunction with external emotional events such as children leaving home and coping with the needs of aging parents or other family members. Many women may develop a mood disorder, as life changes and fluctuating hormone levels combine to cause emotional ups and downs.

Sleep Problems Related to Menopause

More than a third of women in North America today are perimenopausal or postmenopausal. Of those women, about 40–60 percent have sleep problems. A 2003 study reported that the highest rates of sleep problems were found in women whose menopause was caused by removal of the ovaries (48 percent of women with surgical menopause had sleep problems) and those late in perimenopause (45 percent). The rates of sleep problems during menopause were much lower in Japanese women (28 percent) than in Caucasian women (40 percent).

Women who are menopausal or postmenopausal are more likely to have insomnia than when they were premenopausal. A poll conducted on women's sleep in 1998 found that 44 percent of women going through menopause and 28 percent of postmenopausal women have hot flashes at night; on average they had hot flashes three nights a week. This is severe enough to cause trouble sleeping or insomnia for an average of five days each month. Not all groups have hot flashes to the same extent. Another research study from 2003 reported the percentages of women between ages forty and fifty-five who had night sweats: African American women, 36 percent; Hispanic women, 25 percent; Caucasian women, 21 percent; women of Chinese descent, 11 percent; women of Japanese descent, 9 percent.

A 2015 research study showed that women during and after menopausal transition might develop serious insomnia, sleeping on average 43.5 minutes fewer a night during menopausal transition than women not complaining of insomnia. These women were more likely to have hot flashes, and the presence of hot flashes predicted the number of times they awakened during the night. About 50 percent of women with insomnia as they approach menopause sleep less than 6 hours a night.

But not all difficulty in sleeping is caused by hot flashes. Menopausal or postmenopausal women are more likely to have to go to the bathroom at night (43 percent) than premenopausal women (34 percent). Twenty percent of menopausal or postmenopausal women use prescription medications to help them sleep, compared to 8 percent of premenopausal women.

Because smoking and obesity increase the likelihood that a woman will have severe hot flashes, menopausal women should try to stop smoking and bring their weight down. These tasks are obviously a challenge. Menopausal or postmenopausal women on hormone replacement drugs are less likely to have hot flashes during sleep. But because of the perceived risks involved with such medication, many women first try to "sweat it out," not seeking treatment for their hot flashes. The symptoms usually improve with time, as episodes become less frequent and less severe. Women learn what works best for them. Some women dress in layers so that they can remove one or more when a hot flash comes on; others try using lighter bedclothes, sheets, and comforters. Special fabrics such as Sheex, PowerDry, and CoolMax that wick away sweat from the body and help keep athletes dry are now being used for women's bedclothes. Some women find that taking a drink of cold water when a flash begins can lessen the severity, so they keep a large glass of cold water on their nightstand. If none of these strategies helps, or if other symptoms are present, women should consult a doctor.

Medical Treatment of Sleep Problems Related to Menopause

WHAT TO EXPECT FROM THE DOCTOR

In a 2015 study, more than 50 percent of menopausal women reported that their doctor did not seem to recognize the importance of menopause or provided inaccurate information, especially about the use of hormonal therapy. What should women expect from a medical consultation?

Postmenopausal women are at an increased risk for various medical conditions such as heart disease, high blood pressure, osteoporosis (bone thinning), and cancer. A woman seeking medical help should expect her doctor to measure her blood pressure and suggest or conduct the following tests: a Pap smear, tests of blood lipids (cholesterol and triglycerides, which if abnormal increase the risk of heart disease), a breast examination, and often a mammogram. Depending on various risk factors, the doctor might also order bone-density tests since osteoporosis is a common problem in postmenopausal women. A doctor might also order measurements of serum follicle stimulating hormone (FSH) and luteinizing hormone (LH) if it is not clear that menopause has begun. The levels of these two hormones, which change during the normal menstrual cycle, remain elevated when menopause has occurred.

The doctor and patient should discuss risks based on the woman's personal and family medical history. Does the patient have a personal or family history of cancer, stroke, cardiovascular disease, or blood clots in the legs or lungs? The doctor might also order tests to check thyroid status (an overactive thyroid can cause sweating and flushing). All this information can help the patient decide how to deal with her menopausal symptoms.

HORMONE REPLACEMENT THERAPY

Until the summer of 2002, it was widely believed by medical scientists and the public that hormone replacement therapy (HRT) effectively helped prevent cardiovascular disease, osteoporosis, and other disorders that appear to be more common in postmenopausal women than in premenopausal women. But research published in July of that year in the *Journal of the American Medical Association* (*JAMA*) showed that there might not be an overall health benefit to using HRT; there might instead be a greater health risk. The research showed a small (but statistically significant) increase in breast cancer, heart attacks, and stroke among HRT users, but a decrease in colon cancer and fractures. The authors concluded that HRT did not decrease the health risks for the general population of postmenopausal women; in fact, there were more reported heart attacks among women using HRT, though there was no difference in the death rate of women who used HRT over women who did not. A report published in the journal in May 2003 found that in postmenopausal women age sixty-five or older, using HRT composed of estrogen plus a progesterone increased the risk for Alzheimer's disease.

Since then, however, the 2002 study has been reevaluated. Researchers have concluded that the study did not apply to many women who might be using HRT because the subjects in the research study were more than a decade older than most women were at the onset of menopause. We still await a definitive scientific study that can assess the risks and benefits of HRT accurately.

A 2015 report found that the most effective treatment of the vasomotor symptoms of menopause and the treatment that has the greatest effect in improving menopausal women's quality of life is estrogen therapy. Estrogen and progesterone combinations taken for several years can have beneficial effects (fewer bone fractures) or harmful effects (increased risk of breast cancer, gallbladder disease, clots forming in veins, and stroke). Estrogens given alone appear to be safer. They do not appear to increase the risk of developing breast cancer, although there is an increased risk of developing endometrial cancer.

Thus, the best strategy for a woman who has severe sleep difficulties because of menopause is to discuss with her doctor the pros and cons of using HRT medication specific to her medical condition and symptoms. If life with hot flashes is unbearable and results in disrupted sleep with all its related problems, she might find it worthwhile to consider using the medication. Similarly, if her sleep apnea came on abruptly with menopause, she might want to see whether HRT could reverse the problem.

If she decides to use hormone replacement, she should use the lowest effective dose of hormones. A combination of estrogen and progesterone is used to help reduce hot flashes. If the woman has had a hysterectomy she can take estrogen alone.

WHO SHOULD NOT USE HRT

Some women have a disorder that might be made worse by hormone treatments. These women should not use HRT. Certain tumors, for instance, depend on estrogen for growth. These include breast cancer, cancer of the endometrium (lining of the uterus), and melanoma (pigmented cancer of the skin). Women who have had blood clots in their legs, especially if the clots have traveled elsewhere in the body (for example, the lungs), should not use HRT because hormones increase the risk of these dangerous clots.

Women who have a strong family history of one or more of these disorders should discuss using HRT with their doctor. They should have available as much specific information about the medical histories of their blood relatives as possible.

Low doses of antidepressants may be helpful in reducing hot flashes in women who choose not to use hormone replacement therapy. These include venlafaxine (Effexor, Pristiq), paroxetine (Paxil, Pexeva), and fluoxetine (Prozac, Sarafem). The antidepressants are less effective than HRT for severe hot flashes, and side effects may include weight gain, nausea, dizziness, and sexual dysfunction. They can even cause restless legs syndrome (see Chapter 11). These medications are not as effective as estrogen in relieving vasomotor symptoms but result in the largest improvement in psychological well-being.

Clonidine, normally used to lower high blood pressure, may improve hot flashes, but side effects include daytime sleepiness, dizziness, constipation, and dry mouth.

Another prescription medication is gabapentin, an anti-seizure medication that may help in improving hot flashes for women with disturbed nighttime sleep. Side effects may include daytime sleepiness, headaches, and dizziness.

Alternative Treatments for Sleep-Related Menopausal Symptoms

Many women are reluctant to use prescribed medications or HRT. A variety of alternative treatments are available, including traditional Chinese treatments, soy products, and herbal products.

TRADITIONAL CHINESE TREATMENTS

Traditional Chinese treatments include acupuncture, herbal medicines, and moxibustion, in which heated "moxa" (made from the herb mugwort) is applied to the patient. These treatments can offer relief of some symptoms, but no large clinical studies into their effectiveness have been conducted.

SOY AND SOY PRODUCTS

Although there is a widespread belief that a natural substance is safer than a substance that has been manufactured or synthesized, we should not assume that an estrogen-like product made from soy or an herb is intrinsically safer than an estrogen-like product made in a lab or a product like Premarin, produced from chemicals obtained from the urine of pregnant mares. Women who have had breast cancer and are concerned that taking estrogen will increase the risk of a recurrence should be equally wary about taking any substance claiming to have estrogen or estrogen-like properties.

Certain molecules in some plants can affect estrogen receptors in humans. These compounds, called phytoestrogens, are found in soy and are chemically quite different from human-produced estrogens. More than a hundred reports have been published in the medical literature concerning the use of soy in menopause, but their results are contradictory. No detailed reports have been published on how soy products affect sleep disorders, and few of the studies have involved randomized controlled trials, in which the use of a treatment versus a placebo (sugar pill) is compared to determine whether having the treatment is more beneficial than having no treatment.

Hot flashes are reported to be much less common among Japanese women. Some scientists believe that this may be related to the Japanese diet, which includes many soy products. A study from Japan published in 1999 showed that not all soy products were equally effective in reducing symptoms of menopause; it wasn't the amount of soy that was important, but the type of soy. The severity of hot flashes among Japanese women was reduced much more when they ate fermented soy products such as miso, natto, and tempeh than when they ate nonfermented soy products.

Research suggests that using between 30 and 60 grams of soy daily can be effective in reducing hot flashes. A half-cup of a soy product such as miso, natto, or tempeh contains 12 to 16 grams of protein. This means that a person will need to eat a large amount of these products to reach 30 to 60 grams, and gastrointestinal side effects such as gas, bloating, loose stools, and sometimes diarrhea are common.

HERBAL PRODUCTS

Many women prefer to use herbal products for menopausal symptoms, including black cohosh, flax seed, dong quai, jiawei qing'e fang, keishibuku-ryogan, kava, chasteberry, and primrose. Some products may be effective for some populations (for example, keishibukuryogan seems to be more effective for women of Japanese descent).

Studies are continually being reported about their effectiveness, but we still lack well-done long-term studies of the risks and benefits of using most natural products. We thus do not know whether such products are safe when used for years or decades. There are few detailed studies of the effects of such products on sleep problems, and the ones that have been made have reported contradictory results. One study from China showed that black cohosh was effective in lessening hot flashes; another from Thailand found no difference between taking black cohosh and taking a placebo. One study found that 25 percent of preparations labeled cohosh did not contain cohosh!

There is concern that some products, including black cohosh, may cause liver toxicity. Kava may cause a rare but potentially fatal liver problem. The September 2016 *Consumer Reports* lists kava as one of fifteen dangerous supplements to avoid. The U.S. FDA has issued a consumer alert warning people about this problem. Health Canada (the Canadian equivalent of the FDA) concluded that there was insufficient scientific support for the safe use of kava, and that it therefore posed an unacceptable risk to health. In Canada, kava has been recalled from the market. Kava has also been banned in France, the United Kingdom, and Germany. Yet kava has been used for centuries in the South Pacific, so the problems experienced in other countries may lie in how kava is manufactured in the West.

Choosing a Sleep Problem Treatment

After consulting with their doctor or health care provider, and perhaps doing research of their own (a good online resource is the National Institutes of Health website, medlineplus.gov), women seeking to treat sleep problems caused by night sweats and hot flashes should base their decisions on what seems best for them. Some authorities recommend the use of HRT for four years followed by annual reevaluation. The individual woman must decide if the sleep symptoms caused by estrogen deficiency are sufficiently severe and disruptive to warrant treatment. Although such treatment will improve the symptoms, she must consider that she may be increasing her risk of cardiovascular disease and breast cancer. She should also take into consideration that HRT may carry with it the possible benefit of reducing risk of gastrointestinal cancer and problems related to osteoporosis. Each woman should take her own family history into account.

A 2016 review in *JAMA* concluded that no strong scientific evidence supports the use of any natural product for the long-term treatment of menopausal symptoms, and no research has been done on the long-term effects of using these products. At best, the report stated that phytoestrogen products modestly reduce hot flashes but have no effect on night sweats. Until better data are available, women will have to consider with their doctors the risks and potential benefits of such treatments. Many women do use them, and more scientific data should become available as more research is done. It is important, therefore, for women to continue to monitor the results with their doctors.

Disorders Causing Sleeplessness in Postmenopause

Whether they have hot flashes or not, many menopausal women develop insomnia. The reason may be related to the fact that estrogen has many effects on the central nervous system. It is possible that the decrease in estrogen affects the centers of the nervous system that are involved with sleep. In addition, some conditions occur more frequently in older people and, thus, in postmenopausal women. These include:

Mood disorders (discussed in Chapter 16)
Sleep apnea (discussed in Chapter 12)
Movement disorders (discussed in Chapter 11)
Painful conditions, including arthritis (discussed in Chapter 15)
Diabetes (discussed in Chapter 15)
Various cancers (discussed in Chapter 15)

Virtually any condition in this group—and often the medications used to treat them—may be associated with sleep problems. For a woman suffering from one or more of these conditions, the estrogen deficiency of menopause simply complicates an already troublesome part of life.

Perhaps the most troublesome problem is sleep apnea. Although for many years doctors believed that sleep apnea was rare in women, this disorder is actually extremely common, affecting at least 2 percent of adult women. Most women with sleep apnea are postmenopausal; the average woman with sleep apnea is about fifty years old. Harvard University research published in

JAMA in 2003 suggests that by age fifty, women present roughly the same number of new sleep apnea cases as men. Just as estrogen and progesterone before menopause seem to protect women from cardiovascular disease, these hormones also seem to protect women from sleep apnea. Progesterone, which is produced during the menstrual cycle, stimulates breathing, and estrogen is probably responsible for where fat is deposited in a woman's body. When premenopausal women become obese they tend not to have fat deposited in the neck that would increase the risk of apnea.

A recently discovered hormone, leptin, which is produced by fat cells, may stimulate breathing in obese people and may prevent them from developing apnea. This hormone also suppresses appetite. Some people may develop a resistance to the effect of the hormone and so put on more weight and develop apnea.

As a group, women with sleep apnea are older and more overweight as measured by body mass index than are men with the disorder. A Pennsylvania study showed that sleep apnea was much more common among postmenopausal women (2.7 percent) than premenopausal women (less than 1 percent). The same group found that almost all premenopausal women as well as postmenopausal women on hormone replacement therapy with sleep apnea were overweight. Postmenopausal women on HRT had apnea less frequently (0.5 percent) than postmenopausal women not on HRT. Thus, HRT seems to protect against developing sleep apnea. Other researchers have found that apnea is more common in postmenopausal women, affecting perhaps 10 percent of them. In some countries, such as Greece, women diagnosed with apnea were less likely to be obese than American women with the problem.

Women with symptoms suggesting sleep apnea are evaluated exactly as men are, and they receive the same treatments. The focus is on using CPAP, encouraging weight loss, and avoiding alcohol. An obvious question is whether postmenopausal women with apnea should be treated with hormones. In several articles in the medical literature that could be considered pilot studies or reports of individual cases, the authors usually concluded that there may be a benefit in treating postmenopausal women with sleep apnea with hormones (especially estrogen). However, the published results are not highly supportive of the use of HRT to treat sleep apnea in postmenopause because no long-term studies have been done. Large randomized control studies are needed to establish whether HRT works and what constitutes the correct dosage.

Breast Cancer

Breast cancer (which also occurs in men, but less frequently) is a major problem for women before and after menopause, and its treatment can lead to sleep disorders. Because the treatment of breast cancer may result in estrogen deficiency, women who are undergoing or have undergone treatment for breast cancer may have hot flashes that are more severe than those in women going through a normal menopause. In one study, two-thirds of women treated for breast cancer had hot flashes; almost all developed insomnia, and about a third developed a major depressive disorder.

Along with the anxiety of being confronted with a breast cancer diagnosis, the patient still has to face treatment, which often causes great emotional distress. Surgery, loss of a breast and the resulting damage to a woman's self-image, and the use of drugs that may cause acute menopausal symptoms and hair loss can all lead to sleeplessness. Chemotherapy and radiation therapy can also cause insomnia and daytime tiredness.

Breast cancer patients cannot use estrogen-containing medications (HRT) to stop the hot flashes because tumor cells grow more rapidly when exposed to estrogen; thus, these drugs may worsen the woman's prognosis. Many women have tried soy products, but the efficacy and safety of such treatments in breast cancer are not well known. It is worth reemphasizing that we do not know with certainty at this time whether the estrogen-like chemical in plant soy is any safer than estrogen in pill form.

These patients may benefit from some of the treatments suggested for insomnia patients. In particular, seeing a psychologist, pain medications, or sleep-promoting medications may help women sleep during the most difficult periods. Some patients who develop major depression may require psychiatric treatment. Recent research reviewed in 2016 is beginning to focus on improving the daytime fatigue and cognitive deficits ("chemo brain") in these patients with wakefulness-promoting drugs such as modafinil (see Chapter 20).

Andropause

So far in this chapter we have reviewed the sleep issues related to the myriad of changes that occur in women after the end of the reproductive phase of their life. Though sleep problems related to hormone deficiencies are more common among women, some men also experience them when undergoing andropause, the ending of the reproductive phase of a male's life. In contrast to the 100 percent of women who will become menopausal with aging, it is estimated that only about 1–2 percent of men develop a comparable reduction in the production of testosterone, the male sex hormone. The patients may develop anemia (a low blood count), a loss of sexual desire, muscle weakness, and insomnia. Some preliminary studies published in 2015 suggest that these might improve with testosterone replacement. (Astonishingly, there has been almost no in-depth research on sleep issues in these patients.)

But even for men who do not experience andropause, certain conditions can result in an equivalent syndrome: men with advanced prostate cancer or breast cancer who receive therapy to reduce levels of androgens (male sex hormones) can experience similar difficulties to those experienced by menopausal women. The men on such therapy frequently develop hot flashes and night sweats that may disrupt sleep, loss of interest in sexual activity and impotence, breast enlargement and tenderness, thinning of bone, low blood counts, weight gain, and loss of muscle mass. These symptoms may improve with time.

About thirty years ago, studies showed that males who were not producing testosterone might develop obstructive sleep apnea when they started on testosterone replacement. As a result of this early research, doctors are advised to be cautious when starting patients on testosterone who already have sleep apnea, or who have the risk factors for sleep apnea (obesity, for example).

Research published in 2016 suggests that for men with proven testosterone deficiency who have a preexisting cardiovascular disease, testosterone replacement may be useful in preventing future cardiovascular events.

Back to the Insomniac with Night Sweats

I arranged for my patient to have blood tests to rule out excess thyroid secretion as a cause of her symptoms; this disease can cause night sweats (see Chapter 15). She also had a sleep test, which did not demonstrate that she was suffering from restless legs syndrome or dangerous cardiac rhythms during sleep.

Her thyroid function and the amount of iron in her body were normal. She did not have an abnormal heartbeat when she slept or while awake. But her heart rate increased when she woke up during the night. The most impressive part of her sleep study was not the many squiggly lines on her chart, but the video that showed her tossing and turning and constantly changing her position. Even when she slept, she was pulling her blanket off and then on, looking frustrated when she awakened during the night. Following the various tests, it was clear that she had perimenopausal symptoms. Her estrogen levels were dropping, and she was experiencing a sleep problem typical of women in perimenopause. We discussed treatment options including hormone replacement therapy, but she was disturbed by the media stories about its potential dangers and reluctant to use any other medications. She elected to sweat it out.

Hormone levels affect women's sleep throughout their lives. As the levels of sex hormones drop with the onset of menopause, it can trigger a variety of sleep disorders. Doctors are now beginning to understand the relationship between hormones and sleep, and options are opening up to treat these disorders. Hormones are not the only causes of sleep problems, however, and sometimes both doctors and patients have difficulty recognizing when a patient has one of the sleep disorders described elsewhere in this book.

PART TWO

Do I Have a Sleep Problem?

6

How to Identify a Sleep Problem

THE MYSTERY. How can people recognize when they have a sleep problem? And how can they explain the problem so that their doctor will recognize it as well? What are the best words to describe symptoms of sleep disorders to help make sure that you and your doctor are speaking the same language?

~~~~~~~~~~~~~~~~~~~

The Case of the Tired Man with Cancer

There are many times when a doctor thinks, "I wish I had been consulted on this case earlier." This was one of those times. The man in front of me had been complaining of tiredness for about two years. His wife confirmed that he lacked energy. At first his family doctor had assumed that he was depressed. But treatment for depression had not helped, and after routine blood tests revealed nothing unusual, his exhaustion was simply ignored. The patient and his wife were resigned to his being tired, but after about a year, during which the symptom not only continued but became increasingly severe, the wife insisted that the doctor do more tests. One of the tests revealed blood in his stools. This quickly led to the diagnosis of colon cancer, and the patient underwent surgery. The patient had been referred to me because he had insomnia, and his doctor assumed that the insomnia was caused by anxiety over the cancer diagnosis. The patient and the doctor had interacted many times in the two years, but they were not communicating.

Recognizing and Describing Sleepiness

"I am tired." "I feel fatigued." "I have no energy." "I am exhausted." People who think they have a sleep disorder often use phrases like these to describe their symptoms, but their words can mean something quite different to the doctor trying to interpret the symptoms. This miscommunication can stand in the way of a correct diagnosis, and it is often a major obstacle for people seeking help for their sleep problems. For most people, the word *tired* refers to a physical lack of strength, a feeling that it takes too much effort to engage in an activity or remain alert. Someone might feel this way after a hard day of skiing or doing yard work, or if the person had a medical condition that made him or her physically weak. People with lung or heart diseases might also use the word to describe their inability to perform certain activities because of breathlessness. The military and transportation industries will frequently use the word *fatigued* when what they mean is *sleepy*.

To sleep scientists, the word *fatigue* refers to a lack of strength or an inability to perform daily tasks as a result of prolonged activity. The activity could be physical or mental. Patients more commonly use the expression "I have no energy," which is vague. Many people say they have no energy when they actually mean they are very sleepy. Some doctors hearing these words might interpret them to indicate excessive sleepiness and refer the patient to a sleep clinic. But other doctors, hearing these same expressions, might interpret them as signals of depression.

Rather than use these expressions, patients should describe what actually happens. For example, rather than saying, "I am tired," the patient should say, "I always fall asleep watching TV," or "I fall asleep at my computer," or some other specific action. Unless patient and doctor are talking the same language, the doctor might be unable to diagnose the problem.

But how can a person tell if he or she is, in fact, sleepy? By recognizing the signs of sleepiness: someone who has the urge to fall asleep at the wrong time and in the wrong place is sleepy. One commonly used questionnaire to gauge sleepiness in adults is the Epworth Sleepiness Score developed by Dr. Murray Johns at the Epworth Hospital in Australia. Using the form below, patients write down the number (o to 3) that describes how likely they are to fall asleep in different situations. Then they add up the numbers.

People with a total score of more than 12 are as sleepy as people with sleep apnea. People with a score of more than 15 are at high risk of falling asleep when they don't want to. I have seen some sleep apnea patients who score 3 and others who score 24. Of course people who are sleep deprived and are medically normal might have high scores as well.

Some people do not realize that they are sleepy because the sleep disorder has come on gradually. They have become accustomed to feeling sleepy, so they think that what they feel is normal; consequently, they may deny the symptom. Even if people deny feeling sleepy, people suffering from sleepiness can be a hazard behind the wheel or at risk for injury. They need to learn to recognize their sleepiness symptoms.

A patient may be feeling that something is not right, but may be unable or unwilling to confront these feelings or experiences. Yet such experiences may be the direct result of inadequate sleep or a sleep disorder. For instance, a person might complain of having trouble remembering things or concentrating or focusing without realizing that the root of these problems lay in sleepiness. Although the sufferer might not see it, other people often notice when a person is sleepy.

Some people when they are sleepy develop irritability or a personality change. They might become angry easily or for no reason. They might not interact with others at social occasions or might appear uninterested in things. Some people with more severe sleepiness demonstrate automatic behavior during which they perform common activities in unusual ways. For example, someone experiencing a high level of sleepiness might load the dishes into an oven instead of a dishwasher. The person will generally not have any memory of this behavior. I have seen patients who, while in the state of automatic behavior, have committed violent acts against others or their property yet had no memory of the incident.

Modified Epworth Sleepiness Scale

How likely are you to doze off in the following situations?

Use the following scale to choose the **most appropriate number** for each situation:

    0 = would never doze or fall asleep
    1 = slight chance of dozing or falling asleep
    2 = moderate chance of dozing or falling asleep
    3 = high chance of dozing or falling asleep

It is important that you answer each question as best you can.

Situation	Chance of Dozing (0–3)
Sitting and reading (e.g., doctor's waiting room)	_____
Watching TV or using computer	_____
Sitting, inactive in a public place (e.g., a theater or a meeting)	_____
As a passenger in a car for an hour without a break	_____
Lying down to rest in the afternoon when circumstances permit	_____
Sitting and talking to someone (e.g., face to face or on telephone)	_____
Sitting quietly after a lunch without alcohol	_____
Driving a car (e.g., while stopped for a few minutes in traffic)	_____
Add the numbers	_____

Modified from M. W. Johns, "A New Method for Measuring Daytime Sleepiness: The Epworth Sleepiness Scale," *Sleep* 14, no. 6 (1991): 540–45.

Some people experience a form of sleepiness called sleep drunkenness. Such people have much more difficulty becoming alert once they wake up than the average person. Their grogginess might continue for several minutes, up to an hour or more. This is common in severely sleep-deprived people.

A doctor has to sort carefully through the various behaviors described by the patient to discover the sleep deprivation that may be caused by a sleep disorder.

Insomnia

One common phrase used by patients is "I have insomnia." This term is both a symptom and a diagnosis. Simply telling a doctor that one has insomnia or that one is not sleeping is not enough. Is there difficulty falling asleep, staying asleep, waking up too early, awakening unrefreshed? Is the problem one of these symptoms, or several, or even all? What does the patient feel when trying to fall asleep? Is the sufferer's mind racing, or is the problem physical — perhaps the patient cannot stop moving. The more accurately and specifically the patient can describe the symptoms, the more likely the doctor is to diagnose the difficulty. (See Chapter 10 for a fuller discussion.)

Other Sleep Disorders

Being sleepy or being unable to fall asleep is not the only symptom of a sleep problem. There are many types of sleep disorders, and some of them might require treatment or referral to a sleep clinic. The symptoms listed provide clues to these potential problems.

BEING UNABLE TO STOP MOVING

Some patients are fidgety all day and cannot even stop moving at night. Having to keep still (for example, in a movie theater or on an airplane) can be torture. At night the urge to move the legs can become severe, and the only way to relieve the sensation is to move them or to get up and walk around. Some patients complain only about their nighttime symptoms and do not mention the sensation in the legs, instead focusing on the fact that they can't fall asleep. To help the doctor come to a correct diagnosis, they need to describe all the symptoms. They might describe their legs as feeling hot or cold, or note that they feel as though bugs are crawling under their skin. These are all symptoms of restless legs syndrome, which will be discussed in Chapter 11. In some people these sensations may occur in other parts of the body. Even when asleep such patients move a great deal.

DREAMING WHILE NOT QUITE ASLEEP

Some people, in the minutes before falling asleep and sometimes in the minutes directly after falling asleep, have dreams or nightmares that include sounds, visually rich images, or even sensations in various parts of the body. These dreams are called hypnagogic hallucinations. They are not normal. People don't normally dream unless they have been asleep for about ninety minutes. Sleep-deprived people, however, do sometimes have hypnagogic hallucinations, and such symptoms are common in people who have narcolepsy (discussed in Chapter 13). People who experience this type of half-awake hallucination more than once or twice a month should discuss it with their doctor.

FEELING PARALYZED ON AWAKENING

Sometimes people awaken during the night and notice that they cannot move. Usually this is because they have awakened from a dream, although they might not remember the dream. This symptom can last from a few seconds to a few minutes, and it can be frightening, leading to a fear of falling asleep (see Chapter 14). The feeling of paralysis sometimes occurs in people who have no other symptoms of sleep disorder; if the paralysis occurs rarely (that is, once or twice a year), it is generally not considered to be a symptom doctors would treat. But it is also a common symptom of narcolepsy (see Chapter 13). Patients with this symptom who are also sleepy in the daytime may have narcolepsy and should discuss the symptom with a doctor.

DOING WEIRD THINGS WHILE STILL ASLEEP

Sleep behaviors can range from insignificant incidents to actions that might indicate a serious sleep disorder. People who are experiencing the more dangerous symptoms described in this section should discuss them with their doctor.

Sleepwalking and sleep talking. Sleepwalking and sleep talking are very common, especially in children (see Chapter 14). Unless the child does something dangerous, it is not a condition that requires medical attention. Some patients wander and eat while asleep. Often sleepwalkers and sleep wanderers have no recollection in the morning of having done so. When elderly people wander during the evening or at night, it might be a symptom of a more serious problem (see Chapter 7). This behavior is known as sundowning or nocturnal wandering and should be brought to a doctor's attention since it might be a symptom of Alzheimer's disease or related to other factors that could be treatable. Another symptom more common in the elderly is awakening confused and disoriented.

Nightmares. Everyone has had the experience of awakening from a nightmare. This is common, and although it can be very frightening for children — and sometimes for adults — it is not believed to be a symptom of a sleep disorder, unless the nightmares are recurring. If the nightmares are frequently very similar, are violent in nature, and are extremely disturbing, they might be a symptom of posttraumatic stress disorder (see Chapter 16) and should be brought to the doctor's attention. Even sleep terror or night terror — a condition in which people get out of bed during the night or sit up in bed screaming and sometimes sweating, with eyes wide open — is not considered dangerous. It is a variant of sleepwalking. It is sometimes experienced by people with sleep deprivation, however.

Physical reactions to dreams. Some people react physically to or act out their dreams. When this happens, the dreamer might strike out, sometimes hurting his or her bed partner. This is a symptom of REM sleep behavior disorder (RBD), which can be quite severe and dangerous (see Chapter 14) and usually requires treatment. Some people perform sexual acts (sexsomnia) on unwilling partners and cannot remember having done so when they awaken. This symptom should be brought to a physician's attention.

Excessive movements. Some people cannot stop moving when they sleep. They toss, turn, get up and walk around, fidget, and twitch during sleep. Some even exhibit movements as bizarre as simulating riding a bicycle or running. These are symptoms of a possible movement disorder (see Chapter 11). If people experience such behaviors at night and have insomnia or daytime sleepiness, they might require treatment.

Head banging and body rocking. Sometimes people will have a disorder called head banging or body rocking in which they repeatedly bang their head into a mattress or a pillow or even a wall, or move other parts of their body in a repetitive manner. Some patients roll their bodies continuously while sleeping. This is a movement disorder that does not have important health consequences, although it may occasionally result in self-injury.

MAKING NOISES WHILE SLEEPING

Although some of the noises that people make while they are sleeping are not dangerous (for example, the high-pitched grinding sound of bruxism, or teeth grinding, which can be disturbing to the bed partner and can damage the teeth), some can be markers of a severe sleep problem. Sleepers do not hear the noises they make, but the noises can impair the sleep of a bed partner.

Snoring. The most familiar noise made by sleepers is snoring, and the doctor will understand what the patient means by this term. It will be helpful, however, if the patient can describe the snoring as accurately as possible. For instance, snoring can be loud or soft. A bed partner can describe the level and frequency of the patient's snoring to the medical practitioner. The patient or bed partner should try to describe how many nights a week he or she snores, for what proportion of the night, whether the snoring is more common in one position, and whether alcohol makes it worse.

Awful silences. Snoring is usually continuous; it does not change much over the course of a night. Some people, however, have episodes of silence between snores. These silences, particularly if they are long, can be frightening because they indicate that the snorer has stopped breathing. Some patients, especially women with an upper airway resistance syndrome, might have episodes of snorting that be quite brief and are not especially loud. Like periods of silence, snoring and gasps may be a sign of sleep apnea (see Chapter 12) and should also be described to the doctor. At the end of episodes of apnea or the end of the silence, the person might exhibit very loud snorting, gasping, or deep breathing.

Sleep Diaries

A sleep diary can be a useful tool in helping patients document their sleep habits and patterns. It can help patients and doctors identify patterns that may be disturbing sleep and pinpoint a sleep problem. With the help of a sleep diary, a patient might find, for example, that sleepiness commonly occurs after the weekly late evening aerobics class. Parents can recognize whether there is a problem with their child's body clock. The diary might illustrate that during the week, the child falls asleep very late and then sleeps much later on the weekend, behavior that suggests his or her body clock is running late. (The body clock is discussed in Chapter 8.)

The American Academy of Sleep Medicine has a downloadable sleep diary at yoursleep.aasmnet.org/pdf/sleepdiary.pdf. The diary takes only a few minutes each day to complete and should be kept in a convenient place, such as on the bedside table. After the patient has filled in the diary for seven consecutive days or longer, he or she should examine it to see whether any patterns or practices emerge that might indicate symptoms of sleep problems. If there are, the patient should bring the diary when he or she visits the health care provider.

Back to the Tired Man with Cancer

After various tests the tired man with cancer was diagnosed as sleepy, the symptom he had first noticed. But when he described it to the doctor he had used the word *tired*, and the doctor had not asked him specific questions about his sleep patterns, although the insomnia had preceded the cancer diagnosis by about a year. The patient had developed an irresistible urge to move his legs at bedtime, and this interfered with his falling asleep. He had a textbook case of restless legs syndrome, which is often caused by iron deficiency. The patient's RLS came on because he was developing an iron deficiency caused by his slowly bleeding colon cancer. The insomnia and sleepiness were thus both symptoms of the cancer, and had he and the doctor been communicating better, the cancer might have been picked up much sooner. The patient was still iron deficient when I saw him, and I put him on iron medication, which would probably take care of his insomnia.

Patients need to recognize symptoms of their sleep problems so that they will know whether these symptoms indicate a condition requiring medical attention. But equally important is learning how to communicate these symptoms to the doctor. Questionnaires and diaries can help patients and doctors in identifying a sleep problem. Once they learn to speak the same language, the patients and doctors can move forward in dealing effectively with the sleep problem.

7

Secondhand Sleep Problems

THE MYSTERY. When a bed partner snores or makes other noises, such as teeth grinding, or exhibits abnormal sleep behaviors such as sleepwalking (like Lady Macbeth) or restlessness, whose sleep is the most affected?

~~~~~~~~~~~~~~~~~~~~~~~~

The Case of the Snorer's Wife

Patients often come to the sleep clinic with someone accompanying them. Children come with parents, adults with bed partners, and the very elderly with a spouse, a friend, or a child. Sometimes, the doctor's challenge with a new patient is figuring out whether the patient is the one with the sleep problem, such as the time when a husband and wife came into my examination room. The husband took the seat closest to my desk because he considered himself the patient. I asked him a series of questions about his sleep habits, and he answered yes to only one of them: he snored. But his wife had never observed him stop breathing while he slept, he did not feel sleepy during the daytime, he never fell asleep when he did not want to, and he had no trouble staying awake at movies or plays. He had no medical problems that I could discern. His blood pressure was normal. He did not smoke. His caffeine intake was reasonable. He drank alcohol only once in a while, although the snoring was worse on such occasions.

I glanced at the woman and saw that she was looking at the floor. The bags under her eyes were clearly visible. I asked how her husband's snoring was affecting her. After a few minutes, it became apparent that the husband was not the one with the problem. The real patient was the wife.

Who Has the Sleep Problem?

When a bed partner or other family member makes noise or exhibits other sleep-related behavior, it can affect the sleep of others as well. Caring for children, spouses, and elderly parents can leave the caregiver sleep deprived and unable to function effectively. In this chapter I review how sufferers can recognize and manage their own sleep deprivation and their family's sleep problems.

Three universal truths I've discovered about sleeping are:

1. There is perhaps nothing more comforting than watching a loved one sleep comfortably and peacefully.
2. There is perhaps nothing more distressing than observing a loved one's struggle to breathe or battle to achieve restful sleep.
3. There is perhaps nothing more frustrating than having sleep disturbed by a family member whose sleep habits are disruptive but not life threatening.

Coping with the Sleep Problems of Bed Partners

When people share a bed or sleeping quarters, a person with a sleep disorder can severely disrupt the sleep of others. In a barracks, such as a prison or military accommodation, the sleep of many can be disturbed. Most often, though, only a bed partner suffers.

SNORING

Sleepers snore when air attempts to flow through an obstructed breathing passage, making the tissues in the nose and throat vibrate. The sound can vary from a whisper to a noise loud enough to be heard throughout the entire house. The only person who does not hear it is the snorer, who frequently denies that he or she snores. For everyone within earshot, however, the noise can be a form of torture.

I recall interviewing a military officer whose spouse left him to return to her family in Texas. One of the main reasons she left was that she could not stand his snoring and had been unable to get a good night's sleep. People who don't sleep properly are tired, irritable, and snappy; some sufferers in this situation have clinical responses consistent with depression. Such symptoms can affect a marriage.

When both members of a couple snore, a contest seems to develop to see who will fall asleep first. In these snoring relationships, the two keep waking each other up, and after a while one of them usually retreats to another room to snore in peace.

When I treat snoring in a person who does not have a serious health problem such as sleep apnea, what I am really treating is the impact of the noise on the family and others who might suffer as a result of the snoring. Thus, the goal is not to improve the snorer's health, but to help the person most affected by the snoring—the listener! But when the "patient" is not ill, the question arises of how invasive the treatment should be. Should it include surgery, which has risks and may cause a great deal of pain? Can less severe methods lessen the snoring enough to enable family members to get their own needed sleep? The interests of both the snorer and those affected by the snoring must be considered.

The most reasonable approach in this situation is to start with the treatment that is most likely to result in a permanent cure. I counsel snorers to work on reducing their weight (if appropriate), since being overweight often leads to snoring, and to avoid alcohol, which is known to make snoring worse. In fact, snorers should avoid getting into the habit of using any drugs to knock themselves out. Both sleeping pills and alcohol carry the risk of turning snoring into sleep apnea. Sleeping pills and alcohol relax the muscles of the throat that keep the breathing passage open and will worsen breathing. So snorers should avoid alcohol before bedtime.

If the snoring continues, bed partners can consider some behavior modifications to reduce the disturbance to their own sleep. Here are the most common; some have proven more effective than others.

Go to bed first. Some people find that if they fall asleep before the snorer, they are less likely to be awakened by the snoring during the night. Bed partners of snorers might find this effective, but in my experience, it does not tend to work on a regular basis.

Adjust the bed partner's sleep position. People more commonly snore when they sleep flat on their back. Bed partners can encourage snorers to sleep on their side by gently pushing on them or elbowing them in the ribs. Although some sleep experts recommend having snorers sleep with a backpack or a tennis ball sewn to the back of the pajamas (this was the focus of a 2009 research study in sleep apnea patients) to keep them on their side, this approach seems unreliable, and it can cause a sore back. Only 10 percent of patients used the treatment long term. Other gadgets are available to induce people to sleep on their side.

Another option is for the snorer to try sleeping in a comfortable chair in a semi-upright position. When people are in this position they snore less, and even some people with sleep apnea can have more normal sleep when they sit rather than lie down to rest.

Wear earplugs. Research reported in 2012 has shown that male apnea patients are more likely to have hearing loss, a finding that suggests that loud snoring might damage the hearing system. This might also be the case for bed partners of snorers. Earplugs can give both bed partners and family members a better night's sleep; they might also help prevent loss of hearing. Earplugs work extremely well for some people, and there are many different types on the market. They are sold in drugstores, stores that sell industrial and safety clothing, and stores that sell loud machinery such as chainsaws. The earplugs designed for industrial use can be less expensive and might be more effective than some of the earplugs available at other types of retail stores. The wearer should experiment to find the type that fits best, is the most comfortable, and is the least likely to fall out of the ear during the night. It might take a couple of nights for wearers to get used to the earplugs, but the resulting quiet will be worth the effort.

Sleep with a noise machine. Noise machines, which generate "white" noise or other soothing noises, are now readily available to drown out the sound of snoring. In addition to machines, such noises are available as apps for download on iPads, iPhones, and other smartphones.

Sleep in separate beds. Sleeping in separate beds might seem likely to be ineffective, but some snorers also move around a great deal to keep their breathing normal. Thus it is not always the noise that keeps the bed partner awake; sometimes the movements associated with the breathing are more disruptive. Especially if they have slept together for many years, some couples view sleeping in separate beds as a failure of their relationship. But if it enables a normal sleeper to cope with a bed partner's problem, it might actually strengthen the relationship.

Sleep in a separate room. Some sufferers from noisy bed partners leave the bed and find a quieter place to sleep. Ironically, they sometimes do this because they are afraid that their sleeplessness might awaken their snoring bed partner—who is sleeping deeply. Because sleeping on a couch is uncomfortable and unlikely to promote good rest, many sufferers eventually move to another bedroom, often the most distant one from the bed partner.

Buy the property next door. Although I mention this in jest, it has been effective. Since some snoring, particularly at a very low rumbling pitch, can be heard throughout a dwelling, even sleeping in another room might not block out the sound. This was the case with one couple, each of whom snored loudly, who eventually bought the house next door. The two would retire to their respective homes at night. Though drastic, the move saved their sleep— and their marriage.

STOPPING BREATHING

Although listeners can find a bed partner's snoring annoying and disruptive, there is probably nothing more disturbing to them than thinking that their loved one is about to take his or her last breath. Many people cannot get used to sleeping next to someone who stops breathing on a regular basis. They find themselves constantly alert, listening to their partner, waiting to hear him or her take an effective breath. (I review this common problem in depth in Chapter 12.) Sufferers from this kind of sleep disruption should not seek ways to avoid hearing the stopped breathing; rather, they need to insist that the person who stops breathing receives proper medical evaluation. If a patient is diagnosed with sleep apnea and is treated with a continuous positive airway pressure device, bed partners are likely to find the gentle noise of the machine conducive to their own sleep as well.

TOSSING AND TURNING

When one bed partner tosses, turns, gets out of bed, walks around, gets back into bed, and takes half an hour to several hours to fall asleep, the other one is generally feeling angry and frustrated because he or she is not getting any sleep either. Couples may also find that one partner continues moving even after falling asleep, perhaps twitching every twenty or thirty seconds or moving the bedclothes around or kicking. Some partners sweat a great deal. These can be symptoms of a movement disorder (see Chapter 11), and most patients with such a disorder can be treated. If the treatment does not work, separate beds can be a solution.

TEETH GRINDING

Some people grind their teeth while sleeping, a condition known as bruxism. The sound can be one of the most annoying noises for a bed partner to deal with—it sounds as if there is a chipmunk in the bed. Bed partners can try the same techniques suggested for coping with snoring: earplugs, changing bedrooms, and so forth. But bruxism can also indicate a serious dental-health issue: if the teeth grinding becomes severe, it can wear down the teeth to the point that they no longer function well and have to be removed. Teeth grinders should be encouraged to see a dentist, who might fit them with a mouth guard that will minimize the damage that grinding can inflict on the teeth— and their partners' sleep.

TALKING OR WALKING DURING SLEEP

Some people intermittently speak gibberish, moan, or make other strange noises while they sleep. One patient I knew would sit up, sing the national anthem, and then go back to sleep, with no recollection of the event the next morning. These behaviors do not represent anything serious, and in most cases the bed partner becomes used to it. If the bed partner's sleep is continually disturbed, he or she should try the remedies listed above under snoring.

Sleepwalking can also be a disturbance to a bed partner. Some sleepwalkers get out of bed, walk around, even get a snack while sound asleep. They generally return safely to bed and wake up in the morning with no recollection of what they have done. The activity can awaken the bed partner and be a cause of concern. Other sleepers might find themselves waking up in another room of the house. Although most of the time this is not anything to be concerned about because the activities are not dangerous, bed partners disturbed by sleepwalkers can suggest ways to reduce the likelihood of sleepwalking. People are more likely to sleepwalk after they have consumed alcohol and when they are sleep deprived, so sleepwalkers should try to get a lot of rest and avoid alcohol before going to bed. If a bed partner starts doing dangerous things while sleepwalking (leaving the house, cooking on a stove), it is time to seek medical help.

NIGHT TERRORS AND NIGHTMARES

One of the more frightening behaviors for bed partners is when the soundly sleeping person next to them suddenly lets out a bloodcurdling scream and sits bolt upright, pouring with sweat, eyes wide open. Although the bed partner gets a shock, the person experiencing the sleep terrors often retains no memory of what happened. Upsetting as it is for the bed partner, this disorder is not dangerous. It is a variant of sleepwalking and is treated in the same way.

Nightmares, however, when they recur or are particularly violent, can be another matter for both the sleeper and the bed partner. If a sleeper frequently wakes up sweating, sometimes screaming, with heart pounding, and breathing hard, because he or she was frightened by a terrifying dream, especially if the nightmare is recurrent and violent, he or she should be evaluated by a doctor. Nightmares of this type are common among military veterans, who might be suffering from posttraumatic stress disorder (see Chapter 16); such nightmares also occur among people who have suffered various other types of trauma (rape, devastating hurricanes). We now know that PTSD with nightmares can occur after many types of trauma and last for decades.

Patients with REM sleep behavior disorder (covered in more detail in Chapter 14) physically and sometimes aggressively react to their dreams. If they are dreaming that they are being attacked, they might attack their bed partner. They also might punch the walls, throw objects, or jump out of bed, and could hurt themselves and others. This is a serious condition that requires medical treatment.

Coping with Children's Sleep Problems

When children have sleep problems, their caregivers usually lose sleep as well. Children can develop sleep problems at any age, though the types of disorders vary with age. Caregivers who can help their children solve their sleep problems will have a better chance of getting a good night's sleep themselves.

SLEEP PROBLEMS AMONG NEWBORNS

New babies can cause everyone in the house to lose sleep. Until the baby begins to sleep through the night without feedings, the parents, in particular, especially the one who does the night feeding, can expect to suffer sleep deprivation. This is normal. But parents can shorten the period by helping the newborn develop regular sleep habits early. According to Dr. Jodi Mindell of the Children's Hospital of Philadelphia, author of *Sleeping Through the Night: How Infants, Toddlers, and Their Parents Can Get a Good Night's Sleep*, parents can start helping their baby develop positive sleep habits as early as three months of age. These practices will help the baby start sleeping for longer stretches at night. The most important steps are

- developing a regular sleep schedule that is the same every day;
- establishing a consistent bedtime routine; and
- putting the baby to bed drowsy but awake.

Once babies learn to soothe themselves to sleep at bedtime, they will be able to fall back asleep on their own when they wake up at night.

Some problems that arise among babies in their first year are more serious and can have an impact on the entire family. One of the most common is colic; one of the most devastating is sudden infant death syndrome (SIDS).

Colic. At around two weeks of age, about 10 percent of newborn babies start to have unexplained episodes of crying that might occur on a daily basis. Between the episodes of crying, the baby seems fine. These episodes can last for hours and can occur at any time, causing distress for the parents or caregiver and keeping them from getting much-needed sleep. Such behavior is often an indicator of colic. Pediatricians often use the "Rule of 3" to judge whether a baby is colicky: the crying lasts for three hours or more per day, for three days or more per week, for three weeks or longer. Medical science has not found the cause of colic, but some important facts are known. Colic is not caused by gas or abdominal pain. Children with a large amount of gas or problems with excessive diarrhea might be allergic to cow's milk. And it is certainly not the result of bad parenting.

Babies often get over their colic at three or four months, but in the meantime, parents are likely to experience sleep deprivation. Most parents try cuddling and rocking the baby early on, but they should concentrate on trying to normalize their child's sleep habits so that the baby will learn to fall asleep on her or his own.

Establishing a healthy sleep pattern is essential to good sleep. Parents should make sure that their baby wakes up at roughly the same time every morning and goes to bed at roughly the same time every night. Some children who have colic have trouble sleeping even after the colic is gone, a problem that may be related to the parents' not having established a regular sleep schedule.

Some parents have difficulty in tolerating the crying and become angry; they might even violently shake and thus harm the baby. In such situations, parents should take to heart the words of a researcher in 2016: "When a baby won't stop crying despite your efforts, leave and make yourself relaxed first." An excellent source of information about dealing with colic can be found at the Period of Purple Crying website (http://www.purplecrying.info). Although colic can be bad news, the good news is that not every child develops it—and it does go away.

Sudden infant death syndrome (SIDS). Sudden infant death syndrome (SIDS) is the unexpected death of an infant who appeared perfectly healthy. Although the precise mechanisms that lead to SIDS are as yet unknown, SIDS deaths usually occur while the baby is sleeping. One theory holds that babies' nervous systems are not adequately developed so that they are not able to respond when they stop breathing or when their blood oxygen levels dip. SIDS strikes one out of every thousand babies and is more common among babies who are born prematurely and those who are abnormally small at birth. The time of highest risk is when the baby is between two and four months old.

Roughly 90 percent of babies who succumb to SIDS die before they reach six months of age. Research has shown that in the United States, the rate of SIDS might be twice as high for African American babies than for the rest of the population. The reasons behind this discrepancy are unclear, although many scientific articles have reported that it may be the result of socioeconomic factors that lead to babies' being put on unsafe sleep surfaces, sharing beds, or other risky behaviors. Parents can reduce the risk to their child by following such steps as putting the baby to sleep on its back, using pacifiers, and breast-feeding.

A big breakthrough occurred in the 1990s when research showed that babies sleeping in the facedown position were more likely to die of SIDS than those sleeping on their backs. It has been estimated that one-third of SIDS cases are related to sleeping on the stomach. A child sleeping on its stomach might be unable to lift its head if the breathing passage is obstructed, for example, by pillows or blankets. Smoking by the mother before birth and exposure to cigarette smoke after birth have been linked to an increase in the risk of SIDS. Research by a Harvard University group in 2012 showed that secondhand smoke is a risk factor for SIDS: about 20 percent of SIDS cases were probably related to smoking. A study from the United Kingdom showed that alcohol consumption by either parent was also associated with an increased SIDS risk.

Caregivers should follow these 2016 recommendations from the American Academy of Pediatrics to reduce their child's risk of SIDS:

- Place the baby on his or her back on a firm sleep surface such as a crib or bassinet with a tight-fitting sheet.
- Avoid soft bedding, including crib bumpers, blankets, pillows, and soft toys. The crib should be bare.
- Have the baby share a bedroom, but not a bed, with the parents for the first year, or, at the least, for the first six months. (This has been shown to decrease the risk of SIDS by as much as 50 percent.)
- Avoid exposing the baby to cigarette smoke, alcohol, or illicit drugs.

In addition, make sure the baby does not overheat. This can occur if the room is too hot, the baby is wearing too many clothes, or the bedclothes are too heavy, especially when the baby has a fever or an illness such as a cold or other infection. The American Academy of Pediatrics has made available for down-load the 2016 policy statement with detailed recommendations for creating a safe infant sleeping environment (see the Bibliography).

Since the campaign to have children sleep on their backs began, there has been a 40 percent reduction in the number of SIDS deaths. Besides reducing SIDS, sleeping on the back has been reported to reduce episodes of fever, stuffy nose, and ear infections.

CAUTION: Be careful about gadgets that claim to help babies maintain a safe sleeping position. A November 2012 report from the Centers for Disease Control indicated that these may lead to suffocation and death. Parents should check with their pediatrician before using any gadget making such claims.

SLEEP PROBLEMS AMONG YOUNG CHILDREN

Most young children do not have difficulties sleeping regularly. Their parents have established a regular sleeping pattern for them, and they usually sleep through the night. This enables the parents to get a good night's sleep too. But some children have sleep problems that can affect the parents' sleep as well as their own. One of the most common is an inability to fall asleep without a parent.

Children who have learned to associate sleep with being held or rocked may have a great deal of difficulty falling asleep on their own. Some children will crawl into the parents' bed or insist on sleeping in the parents' room. Failure to deal with this problem when the child is young can have long-term implications—I recall a couple who asked me for strategies on how to remove their thirteen-year-old from their bed!

Dr. Richard Ferber of Harvard University discusses this problem in *Solve Your Child's Sleep Problems*, which helps parents teach children to fall asleep on their own. Dr. Ferber's approach works for many families, and I recommend his book highly. The three key steps are:

1. Establish a nightly routine before bedtime that is relaxing to the child. This could include taking a bath, rocking or singing to the child, or ending the day with storytelling or a bedtime book.
2. Put children to bed in their own crib or their own room (if the child has a separate bedroom) while they are still awake but begin-

ning to show signs of sleepiness. This enables children to learn to fall asleep on their own.

3. Leave the room after the child has been put to bed. This is the most difficult part. If the child cries, the parent should wait before going into the room, make the visit brief, and not hold, feed, sing to, or rock the child. The amount of time the parent waits before going into the child's room should become longer and longer on progressive nights. In this way, the child learns that crying brings only a brief visit from the parent, and eventually learns to go to sleep on his or her own. Parents who adhere to the process consistently usually find that within a week or sometimes two the child will establish a regular sleeping schedule. A research study published in 2016 confirmed the effectiveness of this method.

Getting a slightly older child who is sleeping in the parents' bed to move can be more complicated. The child may claim to be frightened to sleep alone. Is "frightened" an excuse or is the child really afraid? One technique that sometimes works when the child is claiming to be frightened is to offer to let the child sleep on the floor next to the parents' bed. After a while, the child will learn that sleeping on the floor is a lot less comfortable than sleeping in his or her own bed. If the child persists in claiming to be frightened, parents should check with a pediatrician to see whether professional help is necessary.

SLEEP PROBLEMS FOUND IN CHILDREN AT ANY AGE
Some children snore, and the causes tend to be different from those of adult snoring. For instance, snoring in children is not usually caused by obesity. Most often snoring or sleep apnea in children is caused by enlarged tonsils and adenoids, or a small jaw.

A child who snores loudly and stops breathing during sleep may have obstructive sleep apnea. If the child has large tonsils or is obese, these problems will probably have to be dealt with. Parents should consult with a doctor about the causes of their child's snoring.

I have also seen many adult snorers whose sleep apnea was caused by an abnormally small jaw; often it turned out that their children also snored and had small jaws. If a parent with a small jaw has sleep apnea, his or her snoring child should be evaluated by a dentist or an orthodontist, as the obstructed breathing pattern may also be due to an abnormal jaw structure. If a child has a very small jaw, orthodontics can often improve the jaw structure and might help reduce the chance of significant sleep apnea years later. The first evaluation visit to an orthodontist is often free.

Other sleep problems in children, such as night terrors, sleep talking, and sleepwalking, are similar to those of adults and can be dealt with in the same ways.

Coping with Sleep Problems Among the Elderly

Many older people enjoy a normal night's sleep. As reviewed in Chapter 2, abnormal sleep is often a sign of a medical condition or the result of a medication or medications. As with sleep problems related to bed partners and children, people who care for elderly ill family members usually experience secondhand sleep problems. Caring for older people who require an intervention at night, such as those who have had a stroke, are on a breathing machine, are incontinent, or require medications to be administered, can have a terrible effect on a caregiver's sleep. Getting help at night from others is critical to maintaining the health of the caregiver.

One of the most common situations depriving caregivers of much-needed sleep is looking after a relative with the neurodegenerative disorder known as Alzheimer's disease (see Chapter 15). More than 70 percent of those with Alzheimer's disease live at home, and they put a great emotional and financial burden on their families and cause many sleepless nights for the caregiver (who is often the patient's daughter). Because the woman is usually the caregiver in her family, she is the one who most often deals with a family member with Alzheimer's disease. Eighty-four percent of caregivers of Alzheimer's patients are women, and on the average the caregivers are sixty-five years old. The average patient lives about eight years after diagnosis. Some researchers believe that the single most important reason why patients with Alzheimer's are institutionalized is that they wander at night and do not sleep. Their caregivers cannot cope with these nighttime activities, which exacerbate their own sleep deprivation. Caregivers of Alzheimer's patients often have to coordinate medical care (medications, doctor visits) and personal care (hygiene, laundry, feeding). Such responsibilities often lead to sleep deprivation.

Caregivers who are experiencing sleep problems should seek help and find out what resources are available. A good place to start is their doctor's office and social service agencies. Getting help from family members can restore sleep and well-being—at least temporarily. Caregivers who are dealing with Alzheimer's patients should not be embarrassed to ask for help.

Back to the Snorer's Wife

The snorer's wife explained how much difficulty she had falling asleep because of the snoring and said that she felt terrible during the day. Her husband retorted that his snoring must not be that loud because it did not wake him up. I asked him whether he would have come to the clinic if his wife had not insisted. He replied no.

It now became apparent to all three of us that although the snoring patient did not have a medical problem, his wife had a serious sleep problem. After discussing different options, the husband agreed to wear a dental appliance that kept him from snoring and allowed his wife to sleep through the night next to him.

Just as patients do not always realize that they have sleep problems, sufferers from secondhand sleep problems might not recognize how a bed partner or family member's sleep habits or disorders can affect their own sleep and health. It can be as important to find treatment for secondhand sleep problems as it is to diagnose sleep disorders in a patient.

8

Resetting the Body Clock

THE MYSTERY. Our brain regulates our body's functions according to an internal circadian clock. When the circadian clock is out of sync with our surroundings, sleep disorders often follow. How can we reset our body clock?

~~~~~~~~~~~~~~~~~~~~~~

The Case of the Night Owl

The thin woman sitting in front of me was in her mid-twenties but looked older. She was haggard, the bags under her eyes were a grayish blue, her eyes were bloodshot, and she had difficulty focusing during the medical interview. I sometimes had to ask her the same question several times. She was obviously exhausted, but she told me that it was taking her three to four hours to fall asleep every night and she could not force herself to get up in time to get to work by eight. Even the several alarm clocks she used didn't help; she slept through them. She had arranged for her mother to give her a wake-up call, but sometimes even the phone's ringing failed to rouse her. Because of her exhaustion, her performance at the office was becoming unsatisfactory, and she felt she was at risk of losing her job.

She told me that she had first become aware of these symptoms when she was a teenager. She would stay up past midnight and many mornings her mother would have to drag her out of bed to go to school. After she got there she fell asleep in her morning classes. Consequently, she had a great deal of difficulty in school, and she barely managed to graduate because she missed so many classes and had poor grades. She was determined to go to college, but she had to take courses in the afternoons to cope with her sleepiness. I asked her whether she ever felt wide awake and alert. She answered, "Weekends. I don't understand it." That was the clue I needed.

Not All Body Clocks Are Set to the Same Time

In Chapter 1, I described the mechanism located in our brains that sleep scientists call the circadian clock. One of the many functions of this clock is to control what time of day we feel sleepy and what time we are alert. In some people, the clock runs late, and as a result, they tend not to become sleepy until late at night. In some people, the clock runs early, and they feel sleepy early in the evening. And in still others, the clock seems to work erratically. These differences in the circadian clock (or circadian rhythms) among the general population are not in themselves symptoms of a disease. People with differing circadian clocks can be perfectly healthy. Problems arise when people's circadian clock is out of sync with the demands of their work or other schedules. Recent research suggests that some of these problems are the result of genetic changes in the system that controls the circadian clock since a genetic component plays a part in determining our circadian rhythms. It is also beginning to appear that abnormalities in the clock might lead to obesity and diabetes.

DELAYED SLEEP PHASE: THE NIGHT OWL

For people with delayed sleep phase (night owls), the clock seems to run three or four hours or more late. It is as though a person who lives in Boston has an internal circadian clock set to Seattle time. Such people do not feel drowsy until around one to three o'clock in the morning or sometimes later. During the week, they have tremendous difficulty getting out of bed to get to work or school, and once there they remain drowsy throughout the day, often nodding off in class or at work. Then on weekends they sleep until noon or even later and wake up feeling alert and refreshed.

Complicating the situation among teenagers is that they often participate in late-night activities involving video games, social media, and texting. It is usually a parent who drags the student with delayed sleep phase out of bed to go to school. Parents can help their children moderate their nighttime behavior so that they can get to sleep at an earlier hour. But rather than sending perpetually sleepy children to school, where they are likely to fall asleep and perform poorly, parents might also need to take their night owl to a doctor.

People who have delayed circadian rhythms frequently go to the doctor complaining of insomnia because they cannot fall asleep at a normal time. Simply put, their internal clock will not let them feel sleepy until it has decided that it is bedtime, and this is as much a physical characteristic as a person's hair or eye color. It is neither normal nor abnormal. Though such people may feel better when they realize they don't have a medical condition, their body clocks can nonetheless cause problems in their daily lives.

Dealing with children or family members with body clock irregularities can be problematic, especially for parents of school-age children. Getting these children through school can be a challenge as it puts stress on the routine that most parents use to keep everyone in the family on schedule.

For children with poor bedtime habits, changes in behavior can result in better sleep habits. Most important, children should limit or eliminate their late-night use of electronics. But when the child is a genuine night owl, parents might try communicating with school officials and teachers regarding their child's physiological need to adhere to a different sleep schedule and the impact this will have on the student's school performance. They can request that schedule adjustments be made to help the student cope. Specifically, they can request that the student be given classes that begin in the afternoon. In my experience, some school systems are accommodating to such requests while others resist them. But parents will find that if the school is willing to accommodate the child's scheduling needs, this can be a good first step in reducing the stress on both student and family.

Probably the best long-term solution for adults with delayed sleep phase is to find a career whose work hours are in sync with their body's schedule. Some of my patients with a delayed clock have been successful in the entertainment business and in service industries such as restaurants because the workday begins later. Today, there are many jobs for people who want to stay up late and sleep in.

Night owls can also experience relationship issues caused by their sleep pattern. Conflict may occur in couples when one partner is an early bird and the other a night owl. When 10:00 P.M. rolls around, one partner might be ready for sleep while the other is ready to rock and roll. Although we might expect night owls to find more success in relationships or marriages with other night owls, it is not always the case. As a colleague noted, "I am a night owl and married to a lark—it works really well having alone time. One daughter is a night owl, the other is a lark."

Although our circadian clocks are to some extent part of our genetic makeup, researchers have found two approaches that can sometimes be effective in changing circadian rhythms. The first, chronotherapy, involves adjusting the sleep schedule so that the sleeper goes to bed two successive hours later every couple of days until he or she has worked his or her way around the clock. Someone whose clock says that a good time to go to bed is 2:00 A.M. will go to bed at 4:00 for two days, then at 6:00 for the following two days, and so on until a desirable bedtime is reached. (Obviously, for schoolchildren, this can only be tried during holidays.)

After the patient has moved the bedtime around the clock until a desirable bedtime is reached, she or he must then consistently choose this time to go to bed. Some sufferers have had success with this method, but they need to be careful, for if they then stay up later for one or more nights, they might fall back into their old routine.

Another treatment involves adjusting the sleeper's exposure to light, which as we saw in Chapter 1 helps regulate the circadian clock. Teenagers, in particular, who use electronic devices at night might find that the light exposure can make it difficult for them to fall asleep, so they need to stop using electronic devices at bedtime. Computer programs and smartphone apps are also available that can adjust the light exposure of device displays.

Morning exposure to bright light from either natural sunlight or a lamp (blue light is the most effective) or even electronic screens can help people who have delayed body clocks wake up. In far northern and southern locations the sun rises late in wintertime; thus, lamps might be used in the mornings to simulate sunlight. Such lamps are used to treat seasonal affective disorder

(SAD), also called winter depression, which seems to be brought on by insufficient sunlight in winter (see Chapter 16). They can also help people regulate
their internal clock. For people with delayed sleep phase, however, such help
might be only temporary: if they stop using the lights, they will probably fall
back into the same undesirable sleep pattern. However, for night owls who
need to be larks, sunlight exposure and chronotherapy are worth trying since
they might offer at least short-term benefits.

Some night owls try to adjust their circadian clocks by using medication.
Melatonin, a sleep-aid supplement billed as the "hormone of sleep," was extremely popular in the 1990s in treating sleeplessness and, in particular, circadian rhythm problems. I am sometimes reluctant to recommend melatonin for
reasons I will address in more detail in Chapter 20. Although it might work for
some people, the long-term implications of its use in children and teenagers
have not been adequately studied.

ADVANCED SLEEP PHASE: THE LARK

Opposed to the night owl is the lark: the person whose circadian clock runs
early. Larks often get tired early in the evening, by 9:00 P.M. or earlier, and are
unable to stay awake longer, then wake up between 4:00 and 6:00 A.M. People
with advanced sleep phase might report symptoms of insomnia and complain
of waking up too early and not being able to fall asleep again. For them it is like
living in Seattle with an internal circadian clock set to Boston time. Such people might seek medical attention because they believe that their early morning
awakenings are abnormal. Since they wake up when everyone else is asleep, it
must be a sign that there is something wrong with them. But as with delayed
sleep phase, advanced sleep phase is not a disease; it is a biological condition
with which some people are born. The best way to cope with a circadian
clock that runs early is to choose a lifestyle and career that is in sync with the
clock.

People with advanced sleep phase usually do not have problems in the
workplace because those with day jobs can arrive on time and stay alert all
day. They are often very successful at certain jobs for which their circadian
clocks represent the ideal schedule. Benjamin Franklin recommended going
to bed early and getting up early. In fact, he thought sleep was a waste of time.
Farmers, surgeons, anesthesiologists, and nurses frequently have the kind of

schedule that suits advanced sleep phase; often they self-select into those careers. In most places of the world, the day shift for nurses begins at 7:00 A.M. and for anesthesiologists and surgeons, the drive from home to work usually occurs before dawn. They are frequently already in the operating room at 7:00. On the other hand, people with advanced sleep phase find night shift work difficult if not impossible and should avoid it.

But although larks have less trouble than night owls coping with standard work schedules, they can have serious difficulties fulfilling social obligations. They are often too sleepy to attend evening events and family get-togethers.

WHEN THERE DOESN'T SEEM TO BE A CLOCK AT ALL

Some people do not seem to become sleepy at night or they sleep at random times over the course of a twenty-four-hour day. It is believed, for example, that Leonardo da Vinci took catnaps every few hours. Parents should be aware and concerned if their teenager starts to have such a sleep pattern; one cause of these problems could be drug use (see Chapter 17). Unlike the advanced and delayed sleep phase patterns mentioned above, this is a disorder (known medically as non-24-hour sleep-wake rhythm disorder) that might require treatment.

This type of abnormal sleep pattern is sometimes seen in people who have suffered damage to the part of the nervous system that controls the circadian clock as well as in people with certain psychiatric disorders. People with certain types of blindness are unable to synchronize their body clocks and might have difficulty sleeping at night. The sleep difficulty may respond to a melatonin-type medication. Although I do not recommend melatonin for people with the owl and lark patterns, I do recommend treatment of patients who have no circadian rhythm if their sleep-wake pattern causes them distress.

AN ENTIRE COUNTRY OF NIGHT OWLS?

Some Americans visiting Spain have discovered that the entire country seems to be on a different clock from theirs. Telephone a restaurant in Spain and ask for a reservation for two at 7:30 P.M. and you are likely to hear the person at the other end suppressing a giggle. You will then politely be told that the restaurant does not open until 8:30 and the chef does not arrive until about 9:00. Restaurants in Spain are packed at midnight. Are culture and lifestyle the only explanation for this?

European Time Zones

As we can see from the time zone map, most of Spain is west of England. Logically Spain should be either in England's time zone or an hour behind England. However, it is not. Spain is in the Central European time zone and has the same time as France, Poland, and Germany. Why? In 1940, Generalissimo Francisco Franco, the dictator of Spain, in an act of solidarity with his ally Nazi Germany, changed Spain's time zone to match that of Germany and therefore the rest of Europe. So on January 1, 2016, sunset was 3:31 P.M. in Chełm, Poland, and 6:13 in Malaga, Spain—a two-hour and forty-two-minute difference, even though the two cities are in the same time zone. Thus days seem to last longer in Spain than in the rest of Europe, and exposure to light keeps people from falling asleep. I am convinced that this is one of the reasons Spaniards are such night owls.

Jet Lag

One of the most common conditions affecting a person's body clock is jet lag. Unknown until the mid-twentieth century—only in the past forty years have people become accustomed to flying very long distances in high-speed air-craft—jet lag can present sleep problems that last for days. Many people find that when they move through several time zones their body clocks become disoriented and confused. (Jet lag does not affect people who fly long distances within a time zone.) Currently, millions of people cross time zones every day. For some it is part of their job. The flight crew of a commercial aircraft is a perfect example. Most have learned to adjust to the jet lag in their daily lives, whereas an ordinary traveler still has to learn.

The impact of jet lag on a passenger's sleep depends on whether the plane is traveling east or west and the number of time zones it crosses, although since the world has twenty-four time zones, a person who crosses twelve is affected equally whether traveling east or west.

FLYING EAST

When people fly east, they "lose" time. On a typical flight from, say, New York to Paris, the flight will take about seven hours. Since Paris time is six hours ahead of New York time, a flight that leaves around 9:30 P.M. will arrive in Paris at 10:30 A.M. When the plane takes off, the body thinks it is 9:30, and, in fact, it is 9:30. But when the plane lands seven hours later, the body thinks it is 4:30 A.M. when it is actually 10:30. Although it is midmorning, the body thinks it is the middle of the night—and you feel as if you have lost six hours. The feeling can be exacerbated by the conditions of the flight. If you are lucky, you will probably sleep for four hours. The first hour in a flight is full of commotion as passengers settle in after take off. The last hour is full of announcements as the plane prepares to land. In the middle of the flight meals are served and movies available. You need to find a way to get your body into sync with local time as quickly as possible so you can make up your lost sleep and function normally on a new schedule.

You should start your adjustments as soon as you get on the plane. Reset your watch to the local time of your destination. Get as much sleep as possible —tell the flight attendants not to offer you dinner or interrupt you for other reasons. Use an eyeshade and earplugs. Do not drink alcohol. Sleeping pills are not recommended for trips of eight hours or less because if a person takes a sleeping pill and is then awakened four to five hours later, the drug may not have worn off. The person could experience a hangover or memory loss, or become disoriented. An article published in the *Journal of the American*

Flying East

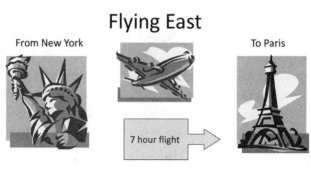

From New York

7 hour flight

To Paris

When you leave your
body thinks it is 9:30
pm and it is 9:30 pm
outside.

When you arrive 7
hours later your body
thinks it is 4:30 am
but it is 10:30 am
outside.

Medical Association in 1987 reported the effects of sleeping pills on three people (neuroscientists, no less!) who had flown to Europe from North America.
Each took a sleeping pill and drank alcohol on the flight. Long after they
had arrived, they realized that none of them had any memory of what they
had seen and done for the ten hours following their landing. Many experienced travelers take melatonin as a sleep aid. (Although many experts recommend against the chronic use of melatonin, they are less concerned about
its occasional use in alleviating jet lag.) One well-respected expert recom-
mends taking 3 milligrams of melatonin at the bedtime of your destination.
If you are flying from North America to Europe in the evening, for example,
take the pill right before takeoff. But make sure the plane will be taking off
on time! If the flight is delayed or canceled, you might find yourself in a very
groggy state.

   After you arrive, adapt your schedule to the schedule at your destination.
If you arrive early in the morning, expose yourself to sunlight, *but not until
your body normally awakens*, about two hours after landing (use sunglasses if
necessary—this will help reset your body clock) and try to avoid taking a long
nap right after you arrive. Doing so may lengthen your adjustment time. Some
flight attendants have told me that when they arrive at their destination, they
go for a workout even though they might be exhausted; the exercise jazzes

Hints for Flying East

Before Takeoff
>   Take 1–3 mg of melatonin at your bedtime at the destination
>   Reset watch

During Flight
>   Wear eyeshades and noise cancelation headphones
>   Take no food or alcohol
>   Sleep

After Landing
>   Avoid bright sunlight immediately after landing
>   Avoid naps
>   Take 1–3 mg melatonin at your bedtime at the destination

them up for a few hours and lets them stay awake through the day and go to sleep at their normal bedtime.

FLYING WEST

Flying east to west presents a different challenge because the traveler "gains" time. A traveler on a short international flight, such as a trip from Paris to New York, would typically arrive in the afternoon, but his or her body would think that it was night. As when traveling west to east, reset your watch to the local time of your destination as soon as you get on the plane. But going east to west on a short flight, try to avoid sleeping for more than a short nap. Eat. Watch the movies. When you arrive, try to adapt your schedule to the schedule at your destination. If you arrive early in the afternoon, don't take a long nap right away. Try to stay awake until your normal bedtime so that you can resume your normal sleep schedule.

A twelve-hour flight, such as San Francisco to Tokyo, offers different challenges. This often leaves between noon and late afternoon, and arrives in the late afternoon or evening. If you changed your watch to Tokyo time, it will seem as though you landed four hours after you took off, except that it is

Flying West, Short Flight

To New York

From Paris

7 hour flight

When you arrive 7 hours later your body thinks it is 8 pm but it is 2 pm outside.

When you leave your body thinks it is 1 pm and it is 1 pm outside.

Flying West, Long Flight

To Tokyo

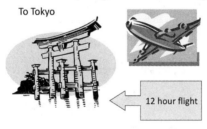

From California

12 hour flight

When you arrive 12 hours later your body thinks it is midnight but it is 4 pm outside. It is also next day.

When you leave your body thinks it is 12 noon and it is noon outside.

the next day. Your body might think it was about midnight when it was only 4:00 P.M.

With the emergence of China as an economic power, there are now many people routinely traveling to China and other countries in Asia. Experienced travelers will try to sleep on westward flights, and many will take a short-acting sleeping pill soon after boarding the aircraft. By the time they land, go through security, pick up their luggage, and get to their hotel, it is night. Usually they are exhausted enough to fall asleep at an appropriate time to wake up at their normal rising time the next morning.

Hints for Flying West

Short Flight (7 hours)

Before Takeoff

Reset watch

During Flight

Stay awake or take a short nap

Eat a normal meal but avoid alcohol

Watch movies, read, or listen to music to stay awake

After Landing

Stay awake until local bedtime

Take 1–3 mg of melatonin at your bedtime at the destination

Long Flight (12 hours)

Before Takeoff

Reset watch

During Flight

Stay awake and eat meal(s) when offered

Avoid alcohol

Watch movies, read, or listen to music to stay awake

Sleep for as long as you can

Eat a meal before landing but avoid alcohol

After Landing

Stay awake until your bedtime at the destination

DEALING WITH JET LAG NO MATTER WHICH DIRECTION YOU TRAVEL

A flight attendant offers the following tips for easing jet lag and adapting to a new time zone whichever direction you travel:

- *Stay awake.* If you arrive at your destination and it's still daylight there, try to resist the temptation to nap. If you must nap, force yourself to get up before the sun sets. Splash cold water onto your face and promptly head outside for some fresh air.

- *Make plans.* Go out with friends—this will give you an incentive to resist cuddling back into that warm and inviting bed. They will be eager to see you. Don't let them down.
- *Exercise.* As counterproductive as it may seem, nothing wakes up a tired body more than a good thirty to forty-five minutes of heart-pumping cardio action. If you work out during the day, when nighttime approaches, you will have a much more restful sleep.
- *Drink lots of water.* And I mean lots. The urge to go to the bathroom often naturally helps you stay awake.
- *Eat light.* But be sure to consume enough to fuel your body. Heavy meals rich in carbohydrates will make you want to sleep. Pass on the pasta, bagels, and ice cream. Eating high-fiber foods such as apples and peanut butter will give you energy and prevent constipation.
- *Drink coffee.* Enjoy as much as you'd like, but if you like it sweet, use a sugar substitute to avoid taking in excessive carbohydrates. Avoid coffee in the few hours before the local bedtime.
- *Have fun.* Do something you love. Staying awake is such a mental activity. Talk on the phone, e-mail a friend, dance, garden, cook, shop, etc. Doing activities you enjoy will distract you from your physical exhaustion.
- *Avoid sedentary activities.* Reading, watching TV, or knitting will most likely make you fall asleep. Stick with activities that involve your body and your mind.

From my own experience, I offer one last piece of advice: If you are traveling to an important meeting at which you must be alert and articulate, do not schedule your arrival for the day of the meeting. About twenty years ago, I attended a medical meeting in Cairns, Australia. For most of the participants, getting there was a long journey. The North Americans flew west. The Europeans flew east. Most participants traveled about twenty to thirty hours. The first meeting was held on the day most of them arrived, and the attendees were exhausted. When the first speaker started showing slides and the lights were dimmed, almost the entire audience fell asleep. Imagine a room full of sleeping sleep specialists who snapped awake when the lights came on between presentations! The following day nobody could remember any of the presentations, although they had probably been excellent. Sleep specialists should know better than to schedule their arrival for the day of the first session!

Returning home in either direction will again result in a shift of the circadian clock. Travelers can count on taking several days to recover from a short trip and up to a week or more to recover from a long trip. Don't schedule anything important for several days after you return.

The Constant Traveler

Some people are constantly on the move; their work does not allow them to settle in any particular time zone for long enough to reset their body clocks to local time. A world-renowned violinist, Leonard Schreiber, described his pattern to me: "I travel constantly crisscrossing the globe and time zones and have to be able to perform at concerts often on the day that I arrive. What works for me is to never try to adjust to local time, but to sleep whenever I am sleepy. I sleep like a cat. My playing is so automatic, that I think that I sometimes play while I am asleep. After a concert I am sometimes so wound up that I can't sleep for hours, and often go out for a bite no matter what the time."

Back to the Night Owl

The patient had classic delayed sleep phase. Her late-running body clock kept her from feeling sleepy so she could not fall asleep until 3:00 or 4:00 in the morning. As a result, during the workweek, she felt as if she were operating in a daze. But on weekends, because she slept until 1:00 or 2:00 in the afternoon, she felt alert and refreshed. Although I explained that her problem was not due to a medical condition, she wanted to try to get herself on a "normal" schedule. The first treatment we tried was chronotherapy. She took a couple of weeks off from work, and every night she went to bed two hours later until she had worked her way around the clock and began falling asleep around midnight. She also went outside as early as she could every morning to expose herself to natural sunlight. At first this worked very well, but after a few weeks she stayed up late at a party one night and her body went back to her original sleep pattern. It was evident that the chronotherapy approach was not going to be effective in the long run.

I again reminded her that she did not have an illness. I advised her to identify with her own slower circadian rhythm just as she does with her hair color and personality. I recommended that she might have a better quality of life if she looked for a job that started in the early afternoon. She took my advice and switched careers. She has for years now successfully worked afternoons and evenings. When last I saw her she was content and married to another night owl. She had adjusted and no longer considered her circadian rhythm a problem.

An ancient link exists between sunlight and our body clocks. Irregularities in a person's circadian rhythm can cause difficulty in falling and staying asleep. Sometimes the irregularity is caused by a body clock that is different from that of most of the population. Sometimes the problem is caused by a body clock that has been confused by travel. But people whose sleep problems are caused by their body clock are usually healthy in spite of their tiredness. They simply have to find the right method of dealing with their situation—whether changing jobs to accommodate their natural sleep schedule or using chronotherapy and sunlight exposure to control it.

9

A World That Never Sleeps

THE MYSTERY. Have humans changed since we conquered the night? Since the invention of the lightbulb, the world never sleeps, but nighttime, irregular-shift, and overtime jobs can contribute to sleep deprivation and health problems.

The Case of the Sleepy Bus Driver

Some of the patients who come to see me at the sleep clinic have complaints that must be addressed immediately. Aircraft pilots, air traffic controllers, ship captains, bus drivers, long-haul truckers, medical personnel, and others whose jobs leave no room for error caused by nodding off constitute emergency cases.

One morning I was consulted by a woman in her early thirties who worked as a public transit bus driver. She was experiencing severe daytime sleepiness and finding it difficult to stay awake on the job. After her supervisor had witnessed her nodding off on a few occasions and threatened disciplinary action, she sought help. We needed to evaluate her to determine whether she had a disease that was making her sleepy and therefore jeopardizing her ability to drive a public bus safely.

To diagnose the problem, I began by asking her the usual medical questions to see whether she had the symptoms of the major sleep disorders, such as sleep apnea, narcolepsy, and severe insomnia, that normally cause excessive sleepiness. She did not. Then I asked her about other possible medical conditions, including thyroid disease and diabetes, which might be associated with her sleepiness. She had no symptoms of those maladies either. But once she started to describe her daily schedule, I knew I'd found the cause of her sleep problem. She didn't have a disease that was making her sleepy and impeding her ability to do her job; it was the job itself that was making her sleepy and thus a danger to the public.

The Graveyard Shift Can Be Hazardous to Your Health

What if you knew that something you were doing might increase your risk for any or all of the following: breast cancer, abnormal menstrual cycles, obesity, abnormal blood lipids, cardiovascular disease? You would want to know what it was, and you would probably want to do something about it. Well, studies have shown that working night shifts puts workers at greater risk than doing day shifts; in fact, links between work schedules and health risks are becoming more and more clear.

Many studies have demonstrated possible connections between shift work and health problems such as cancer. A study from Finland in 2008 found that men working night shift were more likely to develop non-Hodgkin lymphoma, a malignancy involving lymph nodes, than those not working night shift. A study from the Fred Hutchinson Cancer Research Center in Seattle in 2001 reported that women working the graveyard shift (midnight to morning) had a 60 percent higher incidence of breast cancer than those not working this shift. A 1996 study of Norwegian radio and telegraph operators showed that late-shift workers had a 50 percent higher incidence of breast cancer than non-shift workers. A study published in 2012 from the Danish Cancer Society showed that shift work may increase the risk for breast cancer and that the largest risk is associated with the most disruptive shifts (either regular night shifts or rotating night and day shifts). Denmark has awarded compensation to shift workers who have developed breast cancer. And research published in 2016 showed that working at night can increase the risk of estrogen-receptor subtypes of breast cancer, especially in premenopausal women.

A 2011 study from Taipei showed that women working on a rotating shift schedule in semiconductor manufacturing plants were less likely to become pregnant than those on regular shifts and that those who became pregnant were more likely to have smaller babies.

Rotating and night-shift work can also increase the risk of cardiovascular disease compared to day-shift work. A French group examining a decade of scientific studies concluded in 2011 that shift work seemed to affect blood pressure and lipid profile. A 2011 study from Japan reported that the circulation to the heart was impaired in nurses on night shifts. Another, published in 2014, showed that those working night shifts were more likely than those not working nights to develop metabolic syndrome (obesity, hypertension, diabetes), which is linked to heart disease. A 2016 Danish study reported an increased risk of diabetes in nurses who worked the night shift compared to the day shift.

Despite the extant studies, scientists have much to learn about the connections between night-shift work and health. Why, for example, do night shifts increase risks of breast cancer? The connection might lie in the body's timing of hormone secretion. Some hormones, such as melatonin, are secreted primarily during sleep, and recent studies have suggested that melatonin might inhibit certain cancers. A 2016 study from China showed that melatonin levels were lower in women with ovarian cancer compared to healthy women. A person who works at night is exposed to more light than someone who works during the day, and this increased exposure might lead to a reduction in melatonin production. Almost all the cells of the body contain circadian clocks. Our bodies also contain cells called natural killer (NK) cells that kill cancer cells. Studies in 2012 from Rutgers University have shown that disrupting the circadian clock of the NK cells promotes lung cancer growth in experimental animals. All cells contain clock genes and have a circadian rhythm. In 2016, research from Poland found abnormalities of the genes controlling the cell clock of aggressive breast cancer tumors and those lacking estrogen receptors. The graveyard shift may be literally killing workers—and others.

Is Your Pilot Awake? How About Your Doctor?

In today's world, certain jobs carry risks not only to the workers but to all the people their work brings them in contact with. Airline pilots, truck drivers, train operators, bus drivers, and others in the transportation industry; doctors, nurses, anesthesiologists, and others in the health care industry all perform jobs with life-or-death consequences. We put our lives in their hands. We need them to be awake and alert. But the 2012 National Sleep Foundation poll found that most workers in the transportation industries did not get enough sleep. Airline pilots, in particular, drank large amounts of caffeine and took naps to counteract their frequent sleepiness, and they admitted to having made serious errors due to sleepiness. In addition to schedules with long work hours, airline pilots must deal with jet lag, which puts them at greater risk for sleep problems.

No one wants an airline pilot with droopy eyelids. Nor is it pleasant to think that the air traffic controller might be unable to concentrate because he or she is coming off a twenty-four-hour shift. Thousands of jobs require alertness, but workers in the transportation industry have a special need to stay awake. The consequences of their drowsiness can be disastrous. Most transportation industries now have rules about how many hours a person can work and how much time he or she must take off for rest before working again.

Medical personnel are also in the risk business, and sleep deprivation can be a big problem for them as well. A 2011 study from Israel showed that medical residents were more likely to fall asleep while driving or to have automobile accidents after working the night shift. You do not want to be on the road when a sleepy medical resident is heading home from the night shift. Nor do you want a doctor who is unable to focus because she or he spent the entire night in the operating room before your appointment. In the past few years, the ACGME (Accreditation Council of Graduate Medical Education), the organization that accredits medical intern, residency, and fellowship programs, has put in place regulations controlling the number of hours that an intern or resident can work. When I was an intern and then a medical resident, a hundred-hour or more workweek was the norm. Now, interns and residents in hospitals cannot work more than eighty hours a week. A report from Harvard University in 2010 concluded that reducing or eliminating medical resident work shifts of more than sixteen hours resulted in improvements in patient safety and resident quality of life. A report from Stanford University in 2016 showed that patients taken care of by interns and residents who worked more than eighty hours weekly (compared to those working fewer than eighty hours) stayed in hospital longer and were more likely to be transferred to an intensive care unit. At the same time, some controversy remains about whether the eighty-hour-a-week restriction is good or bad since it means there is less continuity of care (the doctors change shifts more often) and less education of the residents (who would spend less time in hospital and thus learn less from their patient experience).

The All-Nighter as a Way of Life

If this book had been written 150 years ago, this chapter probably would not have been necessary. No one had electric lights or telephones, and people worked when it was light and went to bed when it got dark because there was not much else they could do. (It was even difficult to read because it had to be done either by gas lamp or candlelight.) On average, North Americans slept at least an hour more each night than they do today.

Electric light changed all that. In October 1878, Thomas Edison applied for a patent for the electric lightbulb; it was approved in April 1879. The first public demonstration of the lightbulb took place on the last day of 1879. In 1882 the first Americans lost sleep because of electric lights when one square mile of New York City received electric power; there were only fifty-two customers on the first day. But electric utilities soon sprang up all over the country. Electricity came to the White House while Benjamin Harrison was in office (1889–93). In 1892 General Electric, now one of the world's largest companies, was formed, first to generate power and later to produce lightbulbs. By the mid-1930s about 90 percent of urban America and 10 percent of rural America had electricity. Within a decade almost everybody in America had it. Now most of the world's population has access to electricity. The night has been conquered, and the world will never be the same. Electricity and later advances in technology have spurred several trends in the way we conduct our work and social lives, some of which can be contributors to sleep deprivation.

THE TWENTY-FOUR-HOUR WORK WORLD

New York may be the city that never sleeps, but when it comes to work, the whole world never sleeps; in virtually every city of the world there are many people who are working all night in jobs that did not exist in the nineteenth century. Large hospitals never close. Some factories never close. Media outlets (television, print, internet) operate twenty-four hours a day.

In the twentieth century, as electricity and artificial lighting became available, industrialists decided that it would be more productive to keep factories and machinery running round the clock, so they added two extra shifts to each workday. In 1922 Henry Ford introduced the twenty-four-hour assembly line. Every major industry soon adopted his model as a way to increase market share and improve productivity.

The number of industries that never shut down increased; today it is staggering. Because people were now awake, working, and going to and from their homes in the middle of the night, all-night businesses sprang up to service them, including radio and later television stations, grocery stores, and gas stations, as well as all companies to service those businesses, such as the companies that deliver goods to these companies, and companies that offer computer and network tech support. In the twenty-first century, call centers and, later, websites were created to enable customers to make reservations for an airline, find tech support for their computers or dishwashers, or trade on the international commodities market at almost any hour of the day or night. These call centers and industries employ millions of people worldwide, which means that millions of jobs now require their employees to work during the hours normally reserved for sleeping. These working hours can play havoc with the body clock, family life, and, as we have seen, even the workers' health.

A WOMAN'S WORK MAY BE CHANGING, BUT IT IS STILL NEVER DONE

During the world wars women were not allowed to serve in the theaters of war, but they were employed to do the work needed to keep nations running and armaments coming. In addition to more traditional female jobs such as nursing, women now worked in factories, on farms, and in administrative jobs. After the wars ended, many women sought to continue working at these jobs. In the past century women have made enormous strides in the workplace, taking on all the jobs that men do. This includes the high-risk jobs that require special alertness. Women are now long-haul truck drivers, airline pilots, surgeons, police officers, soldiers, and firefighters.

Yet in spite of the strides women have made in the professional world of careers, in most families worldwide, it is the woman who deals with the majority of family and household tasks. Women still are usually the ones who keep track of their family's health issues, know the shoe sizes of their children, schedule the children's after-school lessons and activities. Women still stay up late at night making the children's lunches and cleaning the kitchen.

Although more and more men are taking on these roles, and many families have a single male parent as head of the household, women still take on a disproportionate amount of family care. The number of women in charge of single-parent households in the United States is at least four times greater than that of men, for example. Many women are thus working two jobs: a daytime professional job, and a night and weekend job keeping the household going.

EVEN CHILDREN ARE AFFECTED

Children today have activities available to them that were almost never available to previous generations. When I was growing up, the only after-school activities available to most children were piano lessons and a few afternoon clubs at school. Now some children have access to activities ranging from rhythmic gymnastics to martial arts, debating, chess, and dozens of other hobbies. They take lessons such as ballet, acting, and skating. Even appointments with doctors, dentists, and orthodontists can eat into a child's day. Trying to cram a variety of activities into a twenty-four-hour day can lead to sleep deprivation. Consider ice hockey. If a child is on an ice hockey team, chances are that he or she has played many games that began at 6:00 A.M. because this is when ice time is available. The entire family might leave home at 4:00 or 5:00, in the dark, to take the child to the game. As we have seen, children need on average more sleep than adults, but in today's busy world they are not getting it.

WHAT ALL THIS ACTIVITY IS LEADING TO

The result of these trends is that today's North Americans are overworking, and the more they work, the less they sleep and the worse they feel—and the more likely they are to get sick. A study published in 2003 concerning sleep and health among nurses found that of 71,000 nurses, those sleeping fewer than five hours were 45 percent more likely than those sleeping eight hours to develop a cardiac disease after ten years. The link between sleep and health was also shown in the 2002 National Sleep Foundation Sleep in America poll. This poll of 1,010 adults age eighteen and older found that the less people slept the more likely they were to feel poorly. People sleeping more than six hours on a weekday were more likely to feel optimistic and satisfied with life, while those getting fewer than six hours were more often tired, stressed, sad, or angry.

The same poll found that just over 50 percent of U.S. adults admitted to having driven a motor vehicle while being drowsy in the previous year. Males were much more likely than females to admit to drowsy driving, as were young adults (age eighteen to twenty-nine) and adults with children in the home. Twenty-two percent of males and 12 percent of females admitted to having fallen asleep at the wheel in the previous year!

Today's world is obsessed with speed, communications, productivity, and global competition, and the situation is unlikely to change any time soon. If anything, these trends are getting worse. I have colleagues who spend two to four hours a day commuting to and from work. Even though many of them do their driving in fancy, comfortable cars, math wins the day. A person who spends two hours a day driving to and from work, and works forty-eight weeks a year, spends 480 hours on the road in one year. Some people avoid rush-hour traffic by leaving their homes between 4:30 and 6:00 A.M. and leaving work after 6:00 P.M. In consequence, they hardly ever see their children. They might never see their families in the mornings, and they have only a short time with them in the evenings because they need to go to bed by about 9:00.

So even when people's schedules don't make them sick, the twenty-four-hour workday is affecting their quality of life. But although they cannot change the world, people can manage their work lives to improve their quality of life.

Taking Control of Sleep Time — and Improving the Quality of Life — in Today's Busy World

People have many priorities. In today's world, financial needs (paying the rent or mortgage, finding a job, any job) often take priority over quality of life and health and family. But people whose lives are adversely affected by their work schedule can do something about it. They can start by making a realistic inventory of their values and priorities, then chucking the items that are less important. Once the list has been narrowed down to the two or three most important values, they need to consider whether their lifestyle supports those values. Are they spending time on their priorities? If someone's most impor-tant value is to spend time with the children but his or her schedule makes this virtually impossible, then that person will need to change the schedule. If spending two to three hours a day in a car or working night or other disruptive shifts does not allow someone to do the things that matter most, then he or she needs to find a job that does. If a work schedule leaves someone tired and irritable or is affecting his or her health, it is time to consider changing the schedule or finding a new career. This may be very difficult. (I have used this approach many times with people in their forties who have sleep problems that are related to severe obesity. I remind them that if they remain morbidly obese, and have a severe untreated sleep disorder, they are unlikely to be alive for their children's high school graduation and probably will not live to see any grandchildren. If this is important to them, they need to address the problem and develop a plan to lose weight and treat the sleep disorder.)

SHIFT WORK

An important factor that many people need to consider is whether to take or keep a job with night or irregular shifts. Shift workers need to address the question of how to get the right amount of sleep to keep them healthy, avoid putting themselves or others in danger, and enjoy a productive work and home life. The first step is determining whether their body clocks can adjust to shift work. Remember that people who stay awake nights are generally battling their body clock, which is telling them to go to sleep. It is as though they are subjecting themselves to jet lag every few days. Night owls, on the other hand, are more naturally inclined to do shift work. Their bodies want to stay awake at night. People with late body clocks frequently discover that working night shifts makes a good fit with their life needs. Some people's body clocks even allow them to work rotating or changing shifts without difficulty.

But those for whom staying up late or working at night is ruining their life need to decide whether the advantages of shift work (having a job, earning extra pay) are worth the tradeoff. For instance, the morning host of a radio station in a huge American market used to have to get up at 4:00 A.M. After years of going to bed at 8:30 in the evening and getting up while it was still pitch dark, he decided that his prestigious, well-paying job was not worth the cost. He is much happier now with a less prestigious job with normal hours.

People considering shift work should try to find out what type will work best for them. A fixed shift is one with a regular schedule in which the person always works at the same times. For example, someone might work five nights in a row from midnight to 8:00, have two days off, then work another five nights, and so on. A rotating shift is one in which a person works a few days at one shift, has a few days off, and then switches to another shift. The number of days he or she works in each shift can vary, and the direction of the shifts can vary. One schedule might call for working four nights in a row, then having three days off, then working four day shifts in a row.

A rotation that goes from days (8:00 A.M. to 4:00 P.M.) to evenings (4:00 P.M. to midnight) to nights (midnight to 8:00 A.M.) is a clockwise rotation. A rotation that goes from days (8:00 A.M. to 4:00 P.M.) to nights (midnight to 8:00 A.M.) to evenings (4:00 P.M. to midnight) is a counterclockwise rotation. If there is no pattern or if workers are asked to come in on a random schedule (as is common for long-haul truck drivers), the schedule is referred to as irregular. A split shift is one in which the workday is split by a few hours of no work (more accurately, no pay), usually in the middle of the workday. The worker might be paid for four hours, not paid for four hours, then paid for another four hours. Thus, the worker is being paid for eight hours even though he or she might be away from home for twelve.

In general, fixed schedules are better than rotating schedules for establishing a regular sleep pattern and lifestyle. Rotating schedules are better if the rotations are clockwise. People should avoid work schedules that do not allow sufficient time to deal with family issues or social time. What is the best shift for workers who are also the primary family caregiver, for example? If the children are very young, working an evening or a night shift might make it easier for the caregiver to enlist relatives in helping to care for the children.

HANDLING THE WORKDAY

People should take care not to leave the boring tasks to the end of the day; they might find it difficult to finish. Workers with dangerous jobs or work that incurs risks for others (for example, heavy machinery operators, ships' pilots, air traffic controllers) should make sure that there is a system in place that allows them to be relieved for a rest if they become too sleepy. They might even ask co-workers to warn them when they seem to be losing alertness, and should do the same for others. Spending as much time as possible in well-lighted areas will help keep workers alert, as will eating healthy and nutritious meals and snacks.

Taking naps at appropriate times during the workday can also be a lifesaver for shift workers, especially those on irregular shifts. Workers should investigate whether the employer has a napping policy. A short, fifteen- to thirty-minute nap can dramatically improve alertness for several hours even for a very tired person. Increasingly airlines, for example, are beginning to allow copilots to nap on long-haul flights. The naps must not be too long, however; people who fall into a deep sleep might wake up groggier than they were before the nap. Instead of having doughnuts and coffee during a break, tired workers should consider finding a place to sack out.

Some people will use caffeine in its various forms to try to be alert. Two related products that increase alertness that have been approved by the U.S. Food and Drug Administration for use in "shift-work sleep disorder" are modafinil (Provigil) and armodafinil (Nuvigil).

CHANGING THE SHIFT SCHEDULE

If these strategies are insufficient to help workers cope with their schedule, they should try to change it, even when doing so offers a number of challenges. Some strategies might involve talking with the employer or their union to see what compromises and changes can be made. Workers can make their employer aware of the impact of shift-work schedules on the health and productivity of employees. They can find out whether the work schedules violate labor laws. They should explore the origins of the schedule: Did it come from the employer, the union, a consultant—was it the result of a labor negotiation? In many jobs, a work schedule might have been in place for decades, without anyone giving it much thought. Some companies might be prepared to change schedules based on new scientific information about worker productivity and sleep needs.

Workers might also ask other employees what they think about the schedule, and check to see whether similar industries have better schedules. Before talking to their employer, workers should arm themselves with the appropriate research and any other materials available that can strengthen their case. They should approach the employer not with the problem but with a proposal for a solution, such as a better shift schedule. They might suggest that the company consult with experts to advise them on how to improve schedules. If they present the information in a positive way that would enhance the productivity of the workplace, reduce absenteeism, and improve the health of the workers, they will increase the chances of getting positive action from the employer.

Workers should also educate themselves about what benefits they are entitled to and what protections they have under the law, through the union, or via the employer's own written policies. Most communities operate under federal laws or local regulations that protect workers. Some industries, such as aviation, are closely regulated. Others have regulations that are sometimes loosely enforced—for example, the trucking industry and medical training facilities. One set of regulations may act counter to another set of regulations; in some communities doctors in training were at one time forced to spend more than a hundred hours a week on duty because hospitals argued that these doctors were "training," not "working," which justified their ignoring labor laws. In 2003, and later in 2011, the ACGME, as part of its regulation of medical training programs, started to enforce schedules that do not permit trainees to work more than eighty hours a week. Not all areas are improving, though. At present, a pilot may not be on active flight duty for more than eight to ten hours a shift, but a hospital staff surgeon may operate even if he or she has been awake for thirty-six hours.

AFTER THE SHIFT IS OVER

For workers on the night shift, here are some tips for managing their sleep schedules and staying safe and healthy. First, they should consider forming a carpool. Besides saving on gas, having several people in the car helps the driver to remain alert. Some employers might be prepared to provide taxi transportation home for shift workers if it is not safe for them to drive. If it is sunny outdoors, wearing wraparound sunglasses will help workers who plan to sleep as soon as they get home since exposure to bright light in the mornings might reset their body clock and make it difficult for them to sleep. They should avoid eating heavy meals or drinking alcohol before sleeping. Having a big dinner at breakfast time will make it harder to sleep and might result in weight gain. Drinking alcohol might cause the person to wake up within a few hours. It is not advisable to take sleeping pills when adjusting to or managing a difficult work schedule. Medical science does not know how safe such medications might be over months or years of use.

Workers on night shifts should also make sure that the home environment is conducive to daytime sleep. This might involve installing heavy blinds that darken the bedroom, setting the phone to an answering machine or voicemail, or simply turning off the phone's ringer. In fact, they should turn off all gadgets.

They also need to communicate with family and friends about their sleep requirements so that all will respect their sleep schedule. They might even speak to their neighbors, explaining that they have a job that requires shift work and that they have to sleep during the day. They can politely request that neighbors not make noise that might disturb their sleep. Workers on rotating shifts might even give the neighbors a schedule of which days are affected. Thank-you notes and reminders will usually win their cooperation and understanding. Above all, daytime sleepers should do whatever they need to do to ensure that noisy distractions of any kind are minimized. If loud noise is unavoidable (for people living near an airport or on a bus route, for example), sleepers might find earplugs helpful. Sometimes the only quiet place for daytime sleeping is a remote spot in the basement!

Back to the Sleepy Bus Driver

The bus driver worked a split shift. Every morning, she awoke at 4:00 A.M. to make lunches and get the day organized for her husband and children. At 5:00 she left for the bus terminal and started work at 6:00. Her morning shift was four hours; she finished at 10:00. Between 10:00 and 3:00, she hung around the bus terminal because she did not want to spend two hours going back and forth to her home. Her afternoon schedule was set for rush hour, between 3:00 and 6:00. After she returned to the terminal, she made her way home, arriving at roughly 7:00. Besides being exhausted when she got home, she had little time to divide between her family and sleep. She normally went to sleep at around 10:00, so on average she slept only six hours a night. No wonder she was a wreck and could barely stay awake on the job. After we discussed various options, I contacted the patient's union and employer to see whether anything could be done about her shift schedule since, in my opinion, the schedule was endangering both the driver and the public. I got nowhere. The employer told me that the shifts were controlled by the union. The union head told me that it was too bad that she could not handle the schedule, but better work schedules were reserved for drivers with more seniority, and she would have to work several more years before her schedule would change.

The patient and I discussed the pros and cons of her split shifts, and she decided that her health and her family were more important to her than the job. She decided to change careers. She took another job driving a grocery delivery truck. Her shift ran from 7:00 A.M. to 3:00 P.M. Her income decreased a bit, but she was much happier—and healthier.

Today's twenty-four-hour world might improve productivity, but society could be paying too high a price. Night work, using computers and social media on devices at night, combining professional work with family care, even exploring the range of activities available to children have led to sleep deprivation. One of the most serious results of today's work schedules has been the increase in sleep deprivation among night- and irregular-shift workers. As a sleep specialist, I have encountered an appalling lack of knowledge about the health consequences of shift work by people who were in a position to help their employees, and who should have been motivated to do so, if only to increase their company's productivity and improve customer safety.

PART THREE

Can't Sleep,
Can't Stay
Awake

10

Insomnia

THE MYSTERY. How can one symptom, insomnia, have so many different causes? Can we identify the factors and conditions that may be contributing to the insomnia and help insomniacs get a good night's sleep?

The Case of the Woman Who Refused to Sleep

A patient in her seventies had been referred to me by her family doctor for evaluation of her insomnia. The woman was thin and nervous. She had not combed her hair. She told me it usually took her three to four hours to fall asleep, and once she did she would wake up after an hour or two and be unable to sleep again. She assured me that she did not have continuous movement or restlessness in her legs either at night or in the daytime. Nor was she fidgety. She did not toss and turn; she lay perfectly still in bed whether she slept or not. She did not know whether she snored because she lived alone. When she did fall asleep, she sometimes awakened from a terrible nightmare drenched in perspiration. I asked her whether she was frustrated at not being able to fall asleep quickly, and she said no. Her difficulty sleeping through the night had gone on for so long that she was used to it. When she woke up, she did not have acid in her mouth, heartburn, hunger, chest pain, shortness of breath, or other unpleasant sensations. I was stumped until I asked her how long she'd had the problem and whether she could remember when it started. She told me that she remembered the exact night it started, over twenty years earlier.

What Is Insomnia?

Sleep scientists define insomnia as having at least one of the following symptoms: difficulty falling asleep, difficulty staying asleep, or waking up very early in the morning in conjunction with impaired daytime function. For many people the underlying problem is hyperarousal (the nervous system is jazzed up) even during the daytime; their brains show more metabolic activity in parts of the brain involved with arousal than average in the daytime.

For many people insomnia is not a disease. It is itself a symptom of a disease or other condition. Insomnia is present with many underlying problems, including psychological and psychiatric disorders; medical disorders such as diseases of the heart, lungs, and kidneys; disorders associated with women's life transitions, such as menarche and menopause; and as a side effect of certain medications. Not all people with the conditions just mentioned have insomnia. Scientists believe that the condition along with the hyperaroused brain is what leads to the insomnia and have come up with the term "comorbid insomnia" to define it.

Thus, for people who have trouble falling or staying asleep one of the following might be the underlying cause:

Premenstrual syndrome (see Chapter 3)
Pregnancy (Chapter 4)

Hot flashes (Chapter 5)
Difficulties at home (Chapter 7)
Body clock differences (Chapter 8)
An unusual work schedule (Chapter 9)
Restless legs syndrome (Chapter 11)
Medical conditions (Chapter 15)
Psychiatric conditions (Chapter 16)
Medications (Chapter 17)

In this chapter I review issues that are important to all people with insomnia, focusing on the types of insomnia that are not associated with other
conditions.

How Common Is Insomnia?

Insomnia is not just a problem among overstressed and overworked North
Americans. It is widespread in every country in which it has been studied,
including France, Germany, Great Britain, the Scandinavian countries, Australia, and Japan. In all these places, certain patterns emerge. Insomnia is
common and more frequent among women and older people. One study from
Sweden determined the percentage of thirty-eight-year-old women who had
sleep problems and compared this with the percentage of the same women
who had sleep problems twenty-two and twenty-four years later. The rate of
insomnia had doubled. When they were younger, 17 percent of the women
studied had sleep problems; when the same women were tested later, 35 percent had sleep problems.

Insomnia is also more common among women than men. A U.S. poll
reported in 2002 found that more than half (58 percent) the adult population
suffered from insomnia a few nights a week or more; 63 percent of those reporting insomnia were women. Thirty-five percent of the adults in the sample
reported that they had symptoms of insomnia nightly or almost every night.
According to the survey, insomnia seems to be more common in households
with children, among people who are in poor health, and among shift workers. Compared to people who had insomnia a few nights a week, those people
surveyed who rarely or never had insomnia tended to consider themselves full
of energy, optimistic, happy, relaxed, satisfied with life, and peaceful. People

not getting enough sleep described themselves as likely to become impatient and irritable and to make more mistakes. A 2015 study found that people with painful conditions are much more likely to have insomnia and decreased amount of sleep.

Using the definition of insomnia that includes daytime symptoms (sleepiness, fatigue, poor memory), scientists estimate that about 10 to 15 percent of adults worldwide have chronic insomnia.

When Is Insomnia a Problem?

Sometimes what is perceived as abnormal in fact falls within the normal behavioral range. People sometimes consult a doctor about a symptom that causes them distress but that is not a medical problem. What such people require is reassurance that nothing is wrong. I have had many patients, for example, who find taking thirty minutes to fall asleep distressing, whereas others find it unremarkable. For some, any delay in falling asleep seems endless. Most of the patients that I have seen who suffer from insomnia take more than forty-five minutes to fall asleep.

Over the years, I have seen many patients who have told me that they sleep only six or seven hours a night; they therefore assume they must have insomnia. Many request that I prescribe sleeping pills because they have read that they should be sleeping eight or nine hours a night. I then ask them how they feel during the day and how productive they are. If they say that they feel wide awake and alert and they are functioning at a high level, I tell them that they are among the lucky people who can need less sleep than the rest of the population. Though there are recommendations for the ideal amount of sleep, there is no single number that defines the optimal amount of sleep for each person. Each individual seems to have an optimal number. But although some people recognize how much sleep works best for them, others are unsure of the amount and think they are not getting enough.

If you are uncertain how many hours of sleep you need, try the following experiment. For the next two weeks, record the amount of sleep you have during the night and rate your daytime functioning at the end of the day on a scale from 1 (worst functioning I've ever had) to 10 (best functioning I've ever had). You can rate how productive you are at work, your performance at hobbies and other activities, and how you interact with your family. After two weeks, look at the relationship between your nightly sleep and your next-day functioning. Pay specific attention to whether your sleep amount seems to be highly associated with your next-day functioning. Some people notice, for example, that there are days they function well on relatively little sleep and days they function poorly despite having had a relatively long night's sleep. What most people will learn is that they function better and feel better when they have slept more.

Insomnia Caused by Environmental Factors

Some people have trouble sleeping not because they have a medical problem, but because their environment does not encourage sleep. Too much light in the room can keep people awake or wake them up early in the morning. Good window shades can take care of this. Excessive noise from outside the house (buses, cars, or airplanes) or inside the house (music or television, smartphone notifications) can keep people awake. Wearing earplugs, using noise machines, or turning off the gadgets that ping all night such as smartphones can help. People might also have trouble falling asleep if the temperature is too hot or too cold. This too can be adjusted.

Whenever I travel and stay in a hotel, I always ask for a quiet room facing away from the street, elevators, and ice machines; I also ask for a room without a feather pillow or down comforter. Why no feathers? I learned many years ago, when I was a child, that I was allergic to feathers. Someone who wakes up in the middle of the night with a stuffy nose or sneezing might also be allergic to feathers in the bedding.

People spend on average one-third of their lives in bed—they should make sure that it is not itself the cause of a sleep problem. Is the bed comfortable? Many people have beds that are too firm. Others don't realize that mattresses and even bed frames should be replaced if they are impeding sleep, and they spend their nights trying to sleep on a bed that is lumpy or has a depression in the middle. Another problem is trying to sleep in a bed that is not the right size. Teenagers may suddenly outgrow the standard single bed they have been comfortably sleeping in for many years. Bed partners should take particular care to find a bed in which they will both be comfortable. Both partners may not find the same mattress comfortable and may need to make adjustments. A short "test drive" in the store may not be enough time to determine whether a bed is right; consumers should always make sure before they buy a bed that they can return or exchange it after they have tried sleeping in it. Travelers who find a hotel mattress especially comfortable should ask what brand it is and what model the bed is. Some hotels will sell and ship mattresses to their guests.

People who are really sleepy can sleep almost anywhere, on any surface, but the more comfortable the surface, the better they sleep. A 2012 poll found that the mattress, pillows (on average Americans use two when they sleep), and even fresh scent of sheets all contribute to the quality of people's sleep. But what is true in the United States might not be true in other countries. The ideal conditions for sleeping vary from country to country. In traditional hotels in Japan, for example, a guest might be offered a buckwheat-filled or hard wooden "pillow" and a mat to sleep on.

But just as insomnia might not be a problem in itself but the result of an adjustable environmental factor, it can also be a symptom of a factor that has nothing to do with environment. Two different neighbors of mine came home from a health fair, each with a $6,000 bed. One challenged me to lie in his, and it was certainly comfortable. But did their fancy beds cure their insomnia? No. This is because their insomnia was not caused by their beds. One had severe pain from a back injury, the other was a heavy smoker who would wake up to smoke at night. For them, the insomnia was a different kind of symptom.

Psychophysiological or Learned Insomnia

People who develop insomnia as a result of stress or pain should learn ways to manage or minimize the symptoms; otherwise they risk developing behaviors that perpetuate the insomnia, even after the original cause of the problem has disappeared.

Anyone who has taken a psychology course—and a lot of people who haven't—has heard of Pavlov's dog. In the early twentieth century, the physiologist Ivan Pavlov conducted an experiment in which he rang a bell and then presented a dog with food. Initially the prospect of food made the dog drool; Pavlov would then measure the drool. After repeated sessions in which Pavlov rang the bell and presented the food—reinforcing the dog's drooling—the dog started to drool whenever it heard the bell, even when Pavlov did not give it any food. The dog's behavior showed that it now associated the ringing of the bell with the presentation of food. The significant finding was that the dog had learned this behavior. (This is why a conditioned or learned response to a situation is often called a Pavlovian reflex.)

People also develop conditioned or learned responses. Some people enjoy eating a bag of popcorn when they go to the movies. Over time, if they continue to have popcorn when they go to the movies, they will find that the very act of going to the movies elicits an urge to eat popcorn. For many people, even if they have just eaten a large dinner, the first thing they notice when they walk into a movie theater is the urge to eat a bag of popcorn.

Insomnia can also be learned. Say, for example, that a man suffers a painful back injury that prevents him from falling asleep. In the days and weeks following this injury, as he becomes more and more frustrated about his inability to fall asleep, he begins to associate his bed with that frustration. Even after his injury has healed he suffers from insomnia, yet now it is caused not by the pain but by a new problem, his learned behavior. He now associates the bed with insomnia. No matter how tired he feels when he goes to bed, once he gets into bed, he immediately feels wide awake and alert—the bed itself has become a cue for sleeplessness. And because he gets so little sleep, he attempts to keep alert during the day by drinking large amounts of coffee. But caffeine, especially if taken close to bedtime, can worsen insomnia, thus perpetuating it.

Another common source of learned insomnia is anxiety. Many people have trouble falling asleep before going on a trip. They are worried about whether they will wake up on time or whether there will be other problems, and this anxiety causes them to sleep fitfully. After the trip, if all goes well, everything is fine. But before the next trip, they experience exactly the same anxiety and insomnia until eventually they connect going on a trip with not being able to sleep the night before. When someone expects to have a bad night's sleep, he or she usually experiences a bad night of sleep, and learned insomnia is surprisingly common.

People who experience insomnia very rarely or only at particular times, therefore, probably have nothing to worry about. They can try certain strategies, listed below, to overcome it. But those who have trouble falling asleep most nights and then experience frustration, anxiety, and difficulty in functioning the following day, do have cause for concern. If none of the common cures for insomnia help, they need to bring the problem to the attention of their primary caregiver. He or she, in turn, might direct them to see a specialist in sleep disorders.

Dealing with Insomnia

THE THIRTEEN COMMANDMENTS FOR FIGHTING INSOMNIA

1. Use the bedroom for sleep and sex only.
2. If you can't fall asleep, after fifteen to twenty minutes get out of bed and do something else that is relaxing.
3. Avoid any activity that might cause your brain to become excessively aroused before going to sleep. Avoid arguments, discussions about money or major problems, and exciting television or books. Avoid any vigorous activity for four to five hours before bedtime (sex appears to be the major exception). Turn off all gadget screens an hour before bedtime.
4. Do not consume heavy or spicy meals, which might cause heartburn or discomfort. You want to feel neither full nor hungry before going to bed.
5. If you use an alarm clock, turn it away from you. Do not check the time throughout the night.
6. Establish a relaxing bedtime ritual, such as reading soothing books.
7. If you have nighttime caregiving duties (children, elderly parents, pets), share them.
8. Avoid daytime or evening naps (especially in the four to five hours before bedtime). If you must take a nap, make sure that it is not more than twenty minutes long.
9. Get plenty of exercise but not too close to bedtime.
10. Restrict your time in bed. Spending more time in bed than you need may lead to poor sleep.
11. Take a warm bath or have a hot drink (without alcohol or caffeine) to help you relax.
12. Cut down or eliminate cigarette smoking. Limit caffeine dramatically; if the insomnia is severe, avoid caffeine after lunch. Reduce your consumption of alcohol, which can disrupt sleep.

13. If the insomnia persists, consult your doctor: a medication or a disease might be causing the insomnia.

I discuss other strategies for dealing with insomnia in Chapters 19 and 20.

STRESS AND INSOMNIA

Any type of stress can result in difficulty falling or staying asleep. Examples of situational stress that can cause insomnia include an exam, a meeting, or a trip the next day. Examples of chronic stress might include marital strife, separation, or divorce; financial difficulties; illness in oneself or a family member; or problems in the workplace. Some stresses are out of the sufferer's control and relate to outside factors unconnected to his or her environment. The acute stress Americans felt after the terrorist attacks of September 11, 2001, for example, resulted in a spike in cases of insomnia in the United States. The number of people using medication to help them sleep at least a few nights a month increased from 11 percent of the population to 15 percent. Other examples of this type of stress are unstable international politics that might lead to war or a downward spiral in the national or international economy. Human sleep can be affected by something as intimate as a personal relationship or something as impersonal as a war or witnessing violence thousands of miles away.

For most people, their sleep returns to normal when the factor causing the stress is removed (unless there are psychophysiological factors at work). In a situation in which the stressful situation is expected to be brief but the sleep disruption is severe, medication can be effective. Patients should speak to their doctor about what types of medication are available, the benefits and disadvantages of each, and their costs. (I discuss these medications further in Chapter 20.) Insomnia related to stressful situations is generally expected to improve without treatment, so sufferers should try first to get through their difficult periods without drugs.

In most situations, even when stress occurs over months or years, sleep does return to normal. But there are cases in which the insomnia continues long after the cause of stress has ended. I have seen people in my clinical practice who at one time in their life suffered some enormous stress that was still causing symptoms of insomnia fifty years later. Many Holocaust survivors, to take an extreme case, continue to have trouble falling and staying asleep. Many still suffer from terrible nightmares. Some also suffer from posttraumatic stress disorder, a serious psychiatric condition that is discussed further in Chapter 16.

When a stressful situation is not improving and insomnia continues, the sufferer should seek help to deal with the stress. As a first step, they can look to close friends, family, or clergy. Is there a support system of people they can trust with whom to discuss the problem? Talking with others about a stressful situation might alleviate some of the stress and with it the insomnia. If the situation is severe, sufferers should consider seeking help from a doctor, a counselor, a psychologist, or a social worker.

GETTING HELP FROM A DOCTOR

People seeking medical help for insomnia should not expect the doctor to ask about their sleep habits as part of the routine medical evaluation. This question simply does not come up often enough. The patient needs to bring the issue to the doctor's attention.

The doctor should take the time to discuss the stressful situation and do an assessment to see whether the patient has clinically important depression or a medical problem. What the doctor should not do is write out a prescription for a sleeping pill or an antidepressant without further exploration of the insomnia.

Over the years many people have been referred to me for insomnia after being told by their doctors that they were depressed. Many of these patients were being treated with antidepressants, although some were not depressed. The saddest case I treated was a woman who had developed insomnia on a business trip in the mid-1970s and was started on medications for depression that she was still taking thirty years later. Tragically, the medication she was taking was not even a sleeping pill or an antidepressant; it was an antipsychotic medication that a doctor had prescribed to her in error. She had seen several doctors and had refilled the prescription more than a hundred times, but none of the doctors questioned why she was on this particular medication. The poor woman had been in a fog most of her adult life because of an error in her initial diagnosis and treatment.

When a doctor prescribes a pill to help a patient sleep, it is imperative that the patient ask what the pill is. Is it a sleeping pill or an antidepressant— and has it been approved for use for the treatment of insomnia? Sometimes doctors use medications to treat insomnia that have not been approved for use for insomnia by the U.S. Food and Drug Administration. Since insomnia is a common symptom of depression, antidepressants are frequently prescribed for sufferers. Patients who are prescribed an antidepressant need to ask whether the doctor has diagnosed them as depressed, and if so how long they will be on the medication, and what will constitute a cure.

Patients with insomnia that is not caused by depression and people who are depressed often share a number of common symptoms, such as loss of interest in their usual activities, depressed mood, lack of concentration, reduced memory, fatigue, sleep disturbance, loss of energy, lack of motivation, and irritability. But they usually differ dramatically when it comes to other symptoms. People who are depressed might also suffer from extremely low self-esteem; extreme and inappropriate guilt over past events, wrongdoings, or failings; a high degree of self-blame; and loss of appetite. They might consider suicide.

Patients should not be surprised or concerned if the medical practitioner refers them to another doctor—perhaps a psychiatrist or a psychologist. Most doctors have not received sufficient training on how to deal with the psychologically stressful situations that may lead to sleep disruption. Referral to a colleague is not a weakness on the doctor's part, but rather a strength. It proves that the practitioner knows his or her limitations and can see when it is appropriate to refer the patient to a specialist.

Rare, Incurable Insomnias

TRUE, OR PRIMARY, INSOMNIA

So far in this chapter, I have discussed insomnia as a sleep complaint or a symptom for which sufferers can receive attention and treatment. Some people, however, have what scientists call "true" or "primary" insomnia, and its causes are still a mystery. For 5 to 10 percent of the total population of people who have insomnia, the cause remains unknown. The current theory is that people with this problem are born with an abnormal sleep-generating system.

Most people with this condition have had it all their lives. They frequently claim to have been unable to sleep since they were children; their parents say that they could not sleep even when they were babies. They fussed and had tremendous difficulty falling asleep and sleeping through the night.

People with primary insomnia sometimes have parents or siblings with the same sleep problem, which has led to the theory that this type of insomnia is or can be inherited in some people.

Because medical science does not know the cause of the problem, treatments can only address the symptoms. Some experts have reported good results when patients take a very low dose of antidepressant medication at bedtime. Some patients also do well with hypnotic drugs (sleeping pills).

FATAL FAMILIAL INSOMNIA

Almost no one who reads this book has this problem. It is an extremely rare fatal condition that has been reported mostly in Italy (although a few cases have been reported elsewhere in the world, including countries in Europe and in North America) and has involved only a very small number of families. This disorder is caused by a rare genetic mutation which leads to the production of prions, which are made up of a chemical that is not a virus but behaves like one. The chemical takes over the machinery of the cells to make copies of itself. A study published in 2016 showed that neurons in parts of the brain that normally are involved in regulating sleep are destroyed. Prions also cause mad cow disease (bovine spongiform encephalopathy) and, in humans, Jakob-Creutzfeld disease. Jakob-Creutzfeld causes progressive damage to the nervous system until sufferers also eventually lose their ability to sleep. There is no successful treatment known at this time, and the disease is always fatal. But it is so rare that readers do not have to worry about it—I mention it only to show that insomnia can in extreme cases be associated with death.

Back to the Woman Who Refused to Sleep

When my patient described the night her insomnia started, her story sent a chill down my spine. She had been living in an apartment with her son, who was in his twenties, in a high-crime part of the city. On the night that would change her life forever, two burglars broke down the door to her apartment and killed her son with a knife before they fled. Afterward, the woman became obsessed with personal safety. She tried to stay awake as long as possible every night to make sure no one was breaking into her apartment. The times she woke up in a sweat from a nightmare were terrifying for her because her nightmare was always about the stabbing of her son. After a while, she developed a fear of falling asleep. The woman had posttraumatic stress disorder, which had been untreated for over twenty years. Her insomnia did not bother her much because remaining awake meant that she would not have the awful dreams. I knew I was out of my depth—my training is in internal medicine, pulmonary medicine, intensive care, and sleep disorders. I am not an expert in psychiatric conditions. This patient required expert specialized help. The solution for her was not going to be a sleeping pill but rather a detailed psychiatric assessment and treatment. I referred her to another doctor, and she has apparently done well. As we see in this woman's case, insomnia is often not a disease but rather a symptom of some other disorder. To "cure" the insomnia, the patient or the doctor has to find out what else is going on.

Insomnia, while often distressing, is hard to treat because it has many different causes. Patients and doctors need to isolate the factors perpetuating the insomnia, as well as the comorbid conditions that may be present. Both might require treatment. One such factor is restless legs syndrome, a common cause of insomnia, but one that most doctors don't know anything about.

11

Restless Legs Syndrome

THE MYSTERY. Some people's legs seem to have a mind of their own; they kick and thrash about in bed, keeping their owners (and anyone else in the bed) from sleeping. Yet this common—and often easily treatable—problem is rarely recognized in the medical profession, even among doctors whose family members suffer from it.

~~~~~~~~~~~~~~~~~~~~

The Case of the Doctor's Wife Who Couldn't Stop Moving

At lunch in the hospital cafeteria one of my colleagues asked me whether it would be worth his wife's while to get evaluated for a sleep disorder. He was skeptical because she had not had a good night's sleep for close to thirty years. But after he described the problem, I thought that it was worth looking into.

When I saw the doctor's wife in the sleep clinic, she told a familiar tale.

During her third and last pregnancy, thirty years earlier, she had developed an irresistible urge to move her legs when she tried to go to sleep. She couldn't stop moving. She would toss and turn, but the only way she could find even temporary relief for her discomfort was by getting up and walking around. Because of this it was almost impossible for her to fall or stay asleep. She and her husband no longer shared a bed because the tossing and turning and recurrent awakenings disrupted his sleep as well. Her own interrupted sleep had created the usual daytime problems: lack of energy and a tendency to fall asleep at inconvenient times. She did not want to start on sleeping pills, and no one had been able to suggest another solution. After interviewing and examining her, I ordered some simple blood tests, which gave me the answer to her problem.

The Most Common Medical Problem You Never Heard Of

During pregnancy, many women suffer from a sleep disorder known as restless legs syndrome (RLS), abnormal and excessive movements of the legs while sleeping or trying to sleep. The second most common sleep disorder (after sleep apnea), it affects 15 percent of the general adult population of North America and 10 percent of the female population, yet general practitioners hardly ever diagnose it. It is much more common among older people, affecting 30 to 40 percent of people as they age; it also seems to run in families. It is likely that you or someone you know has this syndrome or will get it. Restless legs syndrome may be caused by various medical conditions, including iron deficiency, but it has other causes as well, and in many cases the cause is never found.

NIGHTTIME SYMPTOMS OF RLS

Most people with RLS complain of insomnia or restlessness at bedtime. Some will experience it in mild form, perhaps repetitive twitching during the night. Some patients say they have "crazy legs." Some patients describe it as an irresistible urge to move their legs, some as a creepy-crawly sensation, either on the surface of the skin or, at times, below the skin. Often patients will say it feels as though insects are crawling under their skin. Others describe severe burning, itching, buzzing, or hot sensations in their feet or legs—when children complain of this sensation, it is often dismissed as "growing pains." (Some RLS patients relate that when they were very young they were considered "squirmy" or restless children and were poor sleepers. The youngest child that I have seen with proven RLS was eight years old.) Most patients find some relief of these symptoms by moving around or walking. For many, the urge to move their feet or to walk around is irresistible. Some people fan their feet at night or apply cold wet towels to their legs. Some patients awaken with painful leg cramps, often in their calves.

Almost all RLS sufferers have trouble falling asleep, and more than three-quarters of them have trouble staying asleep, for even when they are asleep, the twitching does not stop. In roughly 80 percent of people with restless legs syndrome, twitches in their legs (and, less often, in their arms or other parts of their body) occur about every twenty to forty seconds. These twitches can easily be detected in a sleep laboratory. When more than five twitches per hour are present, the patient is diagnosed with periodic limb movement disorder (PLMD). A wide range of movements can occur in patients; some people look as if they are riding a bicycle in their sleep, whereas in other cases only the big toe moves. Sometimes the twitches are so subtle that although the monitors record their electrical activity, observers cannot see any movement. Patients with this milder form are less likely to be diagnosed because the twitching goes unnoticed. These patients need to have an overnight sleep study done.

Bed partners, who often suffer from secondhand sleep disruption, can help identify the patient's RLS, especially if the patient has a movement or a twitch every twenty to forty seconds. Some sleepers with RLS might kick or hit their bed partner. The bed partners of RLS sufferers have told me that sometimes it feels like sleeping beside a wriggling fish, and bed partners might also develop insomnia because of the RLS sufferer's excessive movements. Thus, most RLS patients have two problems: the unpleasant sensations that keep them from falling asleep, and the movements that awaken them or their bed partner. Frequently, couples start to sleep in separate beds and even in separate rooms.

As one woman, who went undiagnosed for forty years, described her experience, "I discovered I couldn't sit still in movies or church and I was walking until three and four in the morning until exhaustion finally let me sleep. I was unable to describe what was happening in my legs. I try to describe the sensation as crawly things under your skin from the knee down, usually starting with a tingly feeling around the knee and then periodic jerks that caused the legs to jump. I have a very loving and sympathetic husband, but eventually we ended up in twin beds. Over the years it gradually got worse, especially with the start of menopause. I feared for what I might do, I got so exhausted and frustrated. I wondered how I could function with so little sleep. I began to panic. My thought at one time was: 'Thank God I don't live in a high-rise—anything to end this.'"

DAYTIME SYMPTOMS OF RLS

Although people complain that their symptoms are the most severe at bedtime, some patients develop an irresistible urge to move their legs or walk around when they are sedentary or in situations that require them to sit still. People with RLS might find it extremely difficult to sit in a car for a long trip; they might need the driver to stop the car frequently so they can get out and walk around. Or they might find it difficult to sit still in a movie theater, irritating the people around them by their constant fidgeting. Even when they are simply sitting in a chair, they might continuously move their legs or tap a heel. As one sufferer described it, "I had to stay still for a bone scan. It was forty-five minutes of sheer torture. Finally they tied my legs down with elastic bands and then had to hold them in place because the spasms were so strong."

Some people with RLS fall asleep at the wrong time or in the wrong place during the day and this is the reason they seek medical help. Their main complaint is not the movements but their severe daytime sleepiness, which can drastically affect their personal and professional life. Even when they fall asleep, the continuous disruptions and movements result in poor sleep quality and sleepiness the next day. Some people find themselves overwhelmingly sleepy and yet unable to sleep. One person who sought help when he was thirty explained that he couldn't sit still in the evening because his right leg would either jerk or get a bone-tickling feeling that forced him to move it. He rarely felt energetic when he got up in the morning no matter how much sleep he had had. For at least twelve years he would fall asleep as soon as he became inactive, no matter where he was; in college he would fall asleep in class and frequently go to the library for the sole purpose of sleeping between classes.

RECOGNIZING RLS IN FAMILY MEMBERS

Often family members can help with a diagnosis of RLS. Bed partners, in particular, are likely to notice movement disorders in their partner. Someone with RLS will sleep fitfully and may get out of bed and return several times a night. He or she might change positions frequently even while sleeping or have repetitive twitches or movements of the legs or, sometimes, the arms. Sometimes they will even kick or hit out. Some RLS patients sweat profusely. If a bed partner shows these symptoms, he or she may have RLS and should bring this to the attention of the family doctor.

We have little data on how common RLS is in young children. A 2015 study found that it is present in probably 2–3 percent of North American children. I have seen it in children as young as eight. One way a parent might recognize a movement disorder in a child is if the child is a very restless sleeper. If he or she tosses, turns, and changes positions constantly, or if the bedclothes are in a tangle in the morning, it might indicate that a movement disorder is present. (It could also indicate that the child has a sleep-breathing disorder, which is discussed in Chapter 12.) If the child also falls asleep in class, it could be another indication that he or she has a movement disorder.

A great deal of research has recently been done on the relationship between attention deficit hyperactivity disorder (ADHD) and restless legs syndrome. For example, one study reported that children who were performing very poorly at school had severe insomnia caused by restless legs syndrome and low iron levels. When the iron levels were corrected, the symptoms vanished and the children's grades improved in school. Parents whose children have been diagnosed with ADHD or ADD (attention deficit disorder) should look into the possibility that the problem might in fact be RLS before putting the child on unnecessary medications.

Restless legs syndrome and periodic limb movements in sleep are extremely common among the older population. The reasons for this are not entirely clear but may in part relate to the fact that many older persons have medical conditions that predispose them to the conditions. Movements may be more common in those with Parkinson's disease, arthritis, anemia, diabetes, and heart disease. As in younger people, difficulty falling asleep and staying asleep may be a clue to a movement disorder.

Causes of RLS

Restless legs syndrome can be caused by a number of factors, which is another reason for getting an evaluation at a sleep clinic. What causes these excessive movements is just beginning to be understood by the sleep medicine community. The impulses that cause the increased movements appear to arise in the central nervous system. And the regularity of the twitches suggests that a pacemaker of some kind within the brain is triggering the movement. Recent research suggests that reduced levels of iron in the brain might play an important role in RLS.

Certain kinds of medication suppress the movements, and this, too, can give doctors an indication of what is causing a patient's RLS. Drugs like ropinirole (Requip) and pramipexole (Mirapex), which increase the amount of dopamine or attach themselves to dopamine receptors in the nervous system, have been successful at treating RLS and suppressing the twitches. Dopamine is one of many chemicals the nervous system uses to send messages between cells. A person with reduced dopamine levels might therefore be more likely to develop RLS, and doctors should look for factors that might reduce dopamine in the central nervous system. For example, iron is involved in the production of dopamine. Reduced iron levels in the body, quite common in women, may be associated with RLS.

For many patients, RLS does not seem to be associated with another medical condition but is rather genetic. (It is quite common among people of French-Canadian descent, for example.) At least six different abnormalities in genes have been linked to RLS, and scientists have shown that RLS (especially if it starts early in life) is an autosomal dominant trait; a person must have one of two abnormal chromosomes to have the trait. Each child receives half of each parent's chromosomes and thus has a 50 percent chance of inheriting this trait from a parent with RLS. The trait has variable penetrance; this means that even if a person carries the gene the symptoms can range from absent or mild to severe. Thus RLS can appear to skip generations. Sometimes families with RLS are found to have low vitamin B12 levels, which could indicate inefficient absorption of B vitamins.

Some medical conditions also lead to RLS, including iron deficiency, anemia, folic acid deficiency, vitamin B12 deficiency, osteoarthritis and rheumatoid arthritis, diabetes, kidney problems, and depression. People with one or more of these medical conditions are at increased risk for RLS and should consult their doctor.

Restless legs syndrome may occur in people with anemia (a reduced level of red blood cells), especially if it is caused by an iron or vitamin B12 deficiency. In fact, recent research has shown that iron deficiency, even when there is no anemia, might still cause RLS. People who donate blood frequently might be at greater risk of developing anemia and restless legs syndrome because when they donate blood they lose iron that might not be replenished by their diet.

Women are at a higher risk than men of developing anemia and iron deficiency because of the repeated loss of blood during the menstrual cycle. This can result in iron deficiency if they do not replace the lost iron in their diet.

Many women first notice restless legs syndrome during pregnancy. About a quarter of pregnant women have RLS by the third trimester. (If RLS is present before pregnancy the symptoms will worsen but may vanish after the birth of the baby.) The fetus is taking iron from the mother, so if her dietary intake of iron does not keep up with what the fetus is taking, she will develop iron deficiency. RLS can also be caused by a deficiency in folic acid, one of the B vitamins, which plays a role in the production of red blood cells. Folic acid deficiency has become less common in North America because of fortification of grains but still occurs. Having adequate folic acid levels during early pregnancy has been linked to a reduced risk of birth defects such as spina bifida.

Because of the importance of B vitamins to red blood cell production, RLS may be more common among people with low levels of vitamin B12. In particular, it might be found among patients (usually older people with pernicious anemia) who do not absorb adequate amounts of B12 from the gastrointestinal tract. Other gastrointestinal disorders such as ulcerative colitis and Crohn's disease might also block the absorption of vitamins or iron and lead to RLS.

Millions of North Americans diet to become or stay thin, and this may result in their not getting enough nutrients from food. A diet deficient in certain nutrients such as iron and vitamin B12 can cause RLS or worsen an already existing case. At the sleep clinic, we have seen RLS caused by low levels of iron in people who avoid red meat or are strict vegetarians.

RLS is also more common in diabetics. This is because of the effect diabetes has on the nervous system. When diabetes has been present for many years, it can damage nerves, resulting in neuropathy, which is believed to be a cause of RLS. About half of all patients with kidney failure who are treated with dialysis have RLS.

At the sleep clinic, we have also found that RLS is more common in people with diseases such as osteoarthritis or rheumatoid arthritis. This disorder has frequently appeared in people waiting for implantation of artificial joints.

We do not yet know the reason these diseases are linked to RLS. The RLS might represent a reaction to the pain caused by these conditions.

Diagnosing RLS

The right treatment for RLS can be prescribed only when the syndrome has been correctly diagnosed. Many doctors do not ask patients about the quality or quantity of their sleep or whether they have symptoms of RLS. People with RLS are too frequently diagnosed with depression and treated with antidepressants. In some cases they might actually be clinically depressed, but often the patients have not been asked about the symptoms of RLS before they are treated for depression. People with severe restless legs syndrome are often very sleepy during the day, and this desire to sleep all the time, along with disturbed sleep at night, can be misinterpreted as a symptom of depression. Compounding the problem, some of the drugs used to treat depression, such as antidepressants, can make RLS worse. At the sleep clinic we have also seen many patients who had been diagnosed as having chronic fatigue syndrome or fibromyalgia who actually had RLS. RLS is perhaps the most common medical condition that doctors fail to diagnose. Patients who suspect they have a movement disorder or are being treated for depression need to talk to their doctor about the possibility that they have RLS and be sure to describe all their symptoms carefully.

When we get a patient at the sleep clinic whom we think might have RLS, we follow a standard sequence to make our diagnosis. First we take a medical history of the patient. This is followed by a complete clinical interview and often a variety of tests.

We expect to find the following symptoms:

- an irresistible urge to move the legs;
- moving the legs or walking lessens the urge;
- worsening of the urge to move at rest;
- worsening in the evening or night.

Less commonly, other muscles (arms or back) may be affected.

If we believe that the patient might have RLS, we order blood tests to see whether anemia is present. Red blood cells contain a pigment called hemoglobin that carries oxygen. When the number of red blood cells or the amount of hemoglobin is too low, anemia is present. Three factors—iron, vitamin B12, and folic acid—are involved in the production of red blood cells and are thus believed to play a role in causing RLS. Iron is found in the red cells and in the bone marrow where the red cells are produced. Iron is also carried by ferritin, the body's major iron-storage protein. The most reliable way to detect iron deficiency, other than examining a bone marrow specimen, is to run a complete blood count and measure the ferritin level.

Even when the blood count is normal, however, the body's iron stores might be reduced. The range of normal ferritin levels is wide, but within the normal range, the lower the ferritin level, the more likely it is that the person is iron deficient. The National Institutes of Health suggest that a ferritin value of less than 75 micrograms per liter in a person with RLS indicates that iron deficiency might be a causative factor. One problem with the ferritin level test is that the level might be elevated when the patient has certain acute or chronic illnesses or inflammatory diseases that mask an iron deficiency.

In some cases the patient might need an overnight sleep study. To monitor movements, activity of the anterior tibialis muscle (the muscle over the shins) is recorded. The sleep study could reveal that the patient takes a long time to fall asleep, tossing, turning, and trying to find a comfortable position.

Patients with RLS show an increase in activity of the anterior tibialis muscles while they are awake. After the patient falls asleep (which may take hours), we frequently detect the repetitive twitches in the muscles, occurring every twenty to forty seconds, symptoms of periodic limb movement disorder.

Although the diagnosis of PLMD is established when there are more than five repetitive twitches per hour of sleep, most patients have many times that number. RLS patients usually have between thirty and a hundred twitches per hour of sleep. When we observe the patients during the sleep test, we can usually see the repetitive movements. But even slight twitches can be detected because each time the patient twitches, his or her pulse (heart rate) increases.

Sometimes patients complain to their physicians about daytime sleepiness; they do not have prominent symptoms of insomnia or restlessness at bedtime but do have repetitive twitches during the night. A sleep clinic evaluation will establish that many of these twitches are linked to brief awakenings of the brain. These short awakenings not only change the brain waves, they also temporarily increase the heart rate. So the quality of sleep decreases, and this could cause sleepiness during the day. People experiencing extreme sleepiness should ask their doctor whether a sleep test is appropriate. The sleepiness could be caused by RLS, although it might also be the result of a coexisting problem such as sleep apnea (see Chapter 12) or narcolepsy (see Chapter 13).

Tests That Can Help in RLS Diagnosis

Complete blood count to check for anemia
Studies (ferritin, serum iron, total iron-binding capacity) to check for
    reduced iron levels
Vitamin B12 level
Folic acid level
Sleep study

Treatment

After RLS is diagnosed, a treatment can be prescribed. The type of treatment will depend on how severe the problem is and what the tests uncover. Sometimes no cause can be determined.

IF A CAUSE OF RLS IS FOUND

At the sleep lab we do not recommend that people treat themselves. Self-treatment without a correct diagnosis can be dangerous. For example, although iron deficiency might be the result of heavy periods or pregnancy, it could also be the result of a serious medical condition such as colon cancer or an inflammatory bowel disease. Taking large amounts of iron if you do not have a severe iron deficiency could lead to serious medical problems.

So if a patient is diagnosed as iron deficient, the cause of the iron deficiency then needs to be determined. If iron deficiency is the cause of RLS, the doctor should prescribe iron, usually in the form of tablets, to be taken for several months. Only about 1 percent of iron taken by mouth is absorbed, so it takes a long time to replenish iron stores in the body. Many preparations or multivitamins contain so little iron that they do not replenish the body's iron stores. The doctor should recommend a preparation that contains the proper amount of absorbable iron. For people, particularly children, who have trouble taking iron tablets, fruit-flavored liquid iron preparations are available. People who become constipated from taking iron supplements could eat prunes; these contain iron and will also help with constipation. In rare cases, injections of iron can be used to treat iron deficiency. The good news is that when treated correctly, improvement can be dramatic. It has been my experience, however, that if the iron deficiency has been present for many years, the RLS might not resolve with iron replacement.

Some people with RLS have a vitamin B12 deficiency because they do not absorb B12 properly from the gastrointestinal tract. These people often require repeated B12 injections.

Women who are pregnant and have an iron or folic acid deficiency should speak to the doctor about proper supplementation. Since red meat is the major source of iron in the Western diet, vegetarians need to find other sources. In particular they need to ensure that they take in enough of the foods (for example, bran flakes, chickpeas, beans, and spinach) that contain significant amounts of iron. Vitamin B12 is found mainly in meat, eggs, and dairy products. Although some plant products contain vitamin B12, they are not as reliable sources of this important vitamin. Vegetarians should consider taking vitamin B12 supplements or eat fortified products to maintain a daily intake of 1.5 micrograms per day (or 2 micrograms during pregnancy).

If the RLS seems to be associated with a condition causing pain, such as arthritis, one management approach is to treat the patient's underlying medical condition. The doctor might prescribe a medication such as a nonsteroidal anti-inflammatory like celecoxib (Celebrex) to treat the arthritis and/or the pain.

Many antidepressants cause RLS symptoms. If the RLS is clearly associated with the use of antidepressant medications, patients should discuss the issue with their doctor, who could suggest alternative treatments. Antidepressants are available that are not associated with restless legs.

IF A CAUSE OF RLS IS NOT FOUND

Frequently doctors cannot determine an obvious cause of a patient's RLS. In such cases, drugs to reduce movements could be prescribed as a treatment option. Three such drugs have been approved for use in the United States by the FDA. Two are medications that are often used to treat Parkinson's disease and one is an anti-epilepsy medication.

Dopamine agonists. Drugs that increase dopamine levels or attach to dopamine receptors in the nervous system have been found to be an effective treatment for RLS. These are the same types of medications used for Parkinson's patients, but at a smaller dosage. Parkinson's disease is also a movement disorder in which dopamine levels are reduced in parts of the brain. The prescription pills pramipexole (Mirapex) and ropinirole (Requip) and the patch rotigotine (Neupro) have been approved for treatment, but patients should take care to follow their doctor's recommendations on how to take them. These medications can have serious side effects, such as daytime sleepiness, worsening of symptoms as the drugs' effect wears off, and, rarely, a tendency toward risk-taking behavior such as gambling.

Anti-epilepsy medications. Another approach, which has become a first-line treatment, is to use drugs that reduce the brain's response to the excessive movements. These drugs permit the brain to ignore or suppress the movements. One drug that has been approved is gabapentin enacarbil (Horizant). Another anti-epilepsy drug that has been prescribed, especially if falling asleep is difficult, is clonazepam, which is in the class of benzodiazepines (see Chapter 20). Sometimes these medications might still be taking effect after the patient wakes up, resulting in a hangover. Those patients might find that taking the medication one or two hours before bedtime helps counter this side effect.

Devices. A device that applies vibrations to the legs (Relaxis) has been approved by the FDA to treat RLS.

For resistant cases. In the most severe cases, when other treatments have not proved effective, we might recommend low doses of codeine or another opiate medication at bedtime, a treatment that has been known to be effective for many years. These are narcotics and are generally classified as controlled substances. How these drugs work is not clear, but they have an effect of decreasing activity in some parts of the nervous system. For example, codeine can suppress the cough center. Patients should discuss this type of medication carefully with their doctor and review the possible side effects.

Other recommendations. Lifestyle modifications can also help decrease RLS. These might include decreasing caffeine and avoiding alcohol and nicotine. Moderate exercise might be helpful, as well as relaxation techniques, massage, and putting hot or cold compresses on the limbs. Some people find that cooling their legs or feet at night helps them; others find the opposite. RLS is truly a mystifying disorder.

Back to the Doctor's Wife Who Couldn't Stop Moving

The patient had restless legs syndrome, brought on by iron deficiency–related anemia. Her tests revealed a very low ferritin level, a low level of hemoglobin, and decreased red blood cells. We did not need to give her a sleep test because her clinical history was so typical of RLS. The treatment we prescribed was to take iron supplements. Three months later, to the astonishment of her husband, her RLS symptoms, which had been present for thirty years, were completely resolved and her sleep became normal.

Many doctors don't know enough or don't ask the right questions about women's sleep disorders, even when the patient is a relative. Restless legs syndrome is a common disorder that causes insomnia. It is particularly problematic in women because it frequently comes on during pregnancy and can also be related to the blood loss of menstrual cycles, which lowers iron levels. It is also extremely common in elderly women. But most patients with RLS can be treated if they can get a correct diagnosis.

12

Sleep Apnea

THE MYSTERY. Many people stop breathing when they sleep, but it was only forty years ago that doctors began recognizing its dangers. Even today, doctors frequently fail to diagnose sleep apnea, a devastating disorder that can play havoc with a sufferer's life—or even kill him or her.

～～～～～～～～

The Case of the Farmer's Daughter

One morning a fourteen-year-old and her father came to see me. She was a slightly overweight blonde with red cheeks who sat and stared at the floor through sad and tired blue eyes. She didn't say much, though I noticed that her mouth was always a little bit open. She had grayish bags under her eyes, which is unusual in teenage girls. Her father, a farmer with overalls and calloused hands, did all the talking.

When I asked why they had come, the father said, "My daughter is a bit slow." I could only imagine the embarrassment and shame this poor child felt hearing those words. He continued to tell me that she had trouble learning and that she was being treated with pills for depression but was not getting any better.

I told the father that I wanted to ask his daughter some questions, and I preferred that she answer for herself. Slowly, the extent of her tragedy emerged. For several years she had been doing poorly in school, and about a year earlier had dropped out because of bad grades. She had had difficulty concentrating and had frequently nodded off in class or fallen asleep during examinations.

I asked the father whether his daughter snored. He said that she had been snoring loudly for several years. I then asked if he had ever noticed that she stopped breathing while she was asleep. The answer was yes.

As I started my physical examination, I knew that I would find the cause of her problem with that simplest of medical instruments, a flashlight.

Sleep-Breathing Disorders

Sleep-breathing problems are so common that almost everyone knows a person who has one. The most common sleep-breathing problems are snoring and sleep apnea. Snoring, the less serious of the two, can cause conflicts in the home because the noise can be disruptive. But sleep apnea, a disorder in which people stop breathing during sleep, can ruin a person's life as it did for my fourteen-year old patient. It can also cause death. Although snoring can be a symptom of sleep apnea, not everyone who snores has sleep apnea.

Snoring

WHEN SNORING IS NOT A MEDICAL PROBLEM

Snoring, the loud noise people make while they are breathing in during sleep, usually signifies that the sleeper's upper breathing passage is obstructed. Snoring results from vibration in tissues as the person tries to suck air in. As we saw in Chapter 6, a sleeper's snoring can be so loud and disruptive that couples have to sleep in different rooms or even on different floors of the house. Couples in my office often get into arguments about whether one of them snores.

The vast majority of people who snore do not have a medical problem. If a snorer experiences no daytime sleepiness and has never been observed to stop breathing, and if his or her blood pressure is normal, then the snoring is not a medical problem (although it may represent a serious problem for the bed partner).

A snorer who has no other sleep-disorder symptoms or medical problems probably does not need an overnight sleep test. But snorers should consider having a medical assessment made, or at least a blood pressure check. In addition to doing a routine check, the doctor might find out whether the snorer's nose is stuffy at night (perhaps the person is allergic to the feather pillow or the cat). The doctor will probably ask whether the snorer has ever had a broken nose or whether he or she wakes up with a sore throat (caused by mouth breathing). The doctor should inspect the patient's breathing passage to make sure that no abnormalities such as a crooked or blocked nose, enlarged tonsils, or other lumps and bumps are present. The doctor should inspect the person's jaw to see whether it is too small or set too far back. The doctor should also check the patient's teeth for overbite. The tongue is attached to the lower jaw. If the jaw is too small or too far back, the tongue will also be too far back, and this can block the breathing passage behind the tongue. This may be a cause of children snoring. These checks may uncover a treatable cause of the snoring, which can improve the quality of sleep for bed partners and others in the household.

Women with no prior history of snoring sometimes start snoring during pregnancy, a symptom that can be related to several factors, including weight gain and the increased hormonal levels during pregnancy. Some women find that their nose becomes stuffed up because they have put on too much weight and the upper breathing passage has decreased in size. And some hormones that increase in pregnancy are believed to relax certain tissues and make the breathing passage floppier. In either case, the airway may become partially blocked and snoring may occur. Most of the time, snoring in pregnancy is not a problem, but the woman might want to discuss the symptom with her doctor to make sure it is not an indication of sleep apnea or a marker of preeclampsia (see Chapter 4). Progesterone, a hormone whose level is high in pregnancy,

besides being involved in reproduction is also a breathing stimulant, protecting the pregnant woman from getting sleep apnea.

TREATMENT OF SNORING

Snoring is not a disease, so I do not recommend any surgical procedures except as a last resort, and only then if the patient has something fixable, such
as a deviated nasal septum. I do not consider surgery necessary as a primary
treatment when there is no medical problem. I have seen many patients over the
years who have had surgical treatment for their snoring, and it has failed to cure
the problem. The type of surgery that is usually performed involves removing
tissue of the soft palate. This will not solve the problem if the obstruction
is somewhere else — for example, behind the tongue — or if the person has a
small jaw. Some treatments that can help snorers who do not have a medical
problem are losing weight, cutting out or down on alcohol and sleep medications, and various dental appliances and other gadgets.

Weight loss. Most snorers are overweight, and the most effective but difficult treatment for those snorers is to lose weight. We have seen dramatic
examples of people who lost weight and cured their snoring. Sometimes a
relatively small weight loss results in a dramatic improvement. Sometimes a
larger weight loss is required.

This is not a diet book, but I will mention a couple of things patients
might want to try that are often not mentioned in diet books. A report published in the *Journal of the American Medical Association* in 2003 on a study of
women over a period of six years found that women who watched two hours of
television a day had a 23 percent increase in risk of becoming obese and a 14
percent increase in risk of developing diabetes. The more they watched, the
greater was their risk. If they watched four hours of television, the risk doubled
to 46 percent for obesity and 28 percent for diabetes. The same study found
that a brisk one-hour walk each day reduced the risk of obesity by 24 percent
and the risk of diabetes by 34 percent.

Avoiding alcohol and sleep medications. One widely recommended treatment for snoring is to avoid alcohol, especially before bedtime. Alcohol, by
reducing the tone of the muscles that keep the upper breathing passage open,
worsens snoring. If possible, snorers should also avoid medications that might
have a similar effect. Such medications might include sleeping pills and other
sleep inducers such as certain types of antihistamines that are used for colds

and allergies. If the label carries a warning about drowsiness, the medication will probably increase the tendency toward snoring.

Dental appliances and other gadgets. People who have a small jaw or an overbite can benefit from using an oral appliance. These appliances resemble the mouthpiece worn by a boxer or football player, but they have to be custom-made to fit the patient's teeth and jaw precisely, and are only worn during sleep. The appliance brings the lower jaw up and forward, and this brings the tongue forward, enlarging the breathing passage behind the tongue. If the cause of the problem is obesity, the appliance is less likely to work.

Many gadgets are available to treat snoring, and many snorers have tried them. Some gadgets work for some people. The most commonly used are adhesive strips placed on the nose to flare the nostrils (Breathe Right) and valves applied to the nostrils (Theravent). Some gadgets are intended to keep the snorer off his or her back. However, I am not aware of any gadgets that are effective enough to be used by all snorers. Sometimes sleeping on the side reduces snoring, so a poke in the ribs by the bed partner can help.

What Is Sleep Apnea?

Sleep apnea is a serious sleep-breathing disorder. The symptoms can be a combination of snoring, pauses in breathing, waking up gasping, and severe daytime sleepiness. Until about two decades ago doctors thought that sleep apnea was primarily a disorder of middle-aged overweight males and that it was nonexistent or rare in females, especially young women. Certainly it was not something we thought a fourteen-year-old girl, such as the one I described in the beginning of the chapter, would have.

The reality is that obstructive sleep apnea is an extremely common condition in both males and females—as prevalent as asthma. The percentage of males with the disorder is roughly double the percentage of females, but it is not rare in females; it affects roughly 2 percent of adult women in North America. It is much more common in postmenopausal women; about 10 percent of postmenopausal women have sleep apnea. Based on 2015 estimates of the U.S. population, this means that there are roughly 4.9 million men and 2.5 million women with sleep apnea in the United States. We now know that obstructive sleep apnea can strike people of all ages, from newborns to adults in their nineties. And not all sufferers are overweight.

The stereotype of the middle-aged male victim of sleep apnea has resulted in a frequent failure to recognize the disorder in women and children. Not only were far fewer women with sleep apnea diagnosed and treated, but women with sleep apnea were frequently treated for the wrong condition. We found, for example, that women with sleep apnea were frequently treated for depression. One twenty-nine-year-old woman explained: "I was having anxiety attacks, sleepiness, and depression. I spoke with my doctor and she suggested that I begin taking an antidepressant. My doctor and I had tried another drug the year before, but neither was working for me. I started thinking that I was becoming overstressed at my job, and I wasn't able to cope on some days without becoming ridiculously irritable. . . . I found I was in bed sleeping a lot and was never feeling refreshed. I was gaining weight and not feeling any better."

Research from Harvard published in *JAMA* in 2003 shows that at age thirty men are five times more likely to develop sleep apnea over a ten-year period than are women. By age fifty the tables have turned; at that point, women are two times more likely to develop sleep apnea over a ten-year period than are men.

Women are also more likely to have a variant of sleep apnea called upper airway resistance syndrome (UARS), though the percentage of women with UARS is currently not known. Children may develop sleep apnea because they are born with abnormalities involving their face (for example, a small jaw) or enlarged tonsils and adenoids or because they are obese.

Even your dog might have sleep apnea. Sleep apnea occurs in several breeds of dog, especially those with a "flat" face (a short snout), such as bulldogs, boxers, and Boston terriers.

THE "DISCOVERY" OF SLEEP APNEA

What happened to people with sleep-breathing disorders before the 1970s? Doctors usually have some patients whose clinical history is permanently etched in their memory. I encountered one such patient when I was a medical intern at the Michael Reese Hospital in Chicago between July 1971 and 1972. (This hospital has since closed down.)

I was on a rotation on one of the medical wards, and one of the patients, a woman, had me completely stumped. The medical staff at the hospital was trying to determine the cause of her severe, incapacitating sleepiness. Almost every time I went to see her, she was sound asleep, and when she woke up, she still seemed sleepy. Yet in all other respects her health was normal. However, she

The Fat Boy (detail), Charles Dickens, *The Posthumous Papers of the Pickwick Club* (1836)

was about a hundred pounds overweight, and this combined with the sleepiness indicated to me that she might have something I had learned about in medical school called the Pickwickian syndrome. In his first novel, *The Pickwick Papers* (1836), Charles Dickens had described the features of this condition in a character called Joe the Fat Boy, who snored and was constantly sleepy. In his first appearance: Joe is described as "a fat and red-faced boy, in a state of somnolency."

I had learned that the Pickwickian syndrome was a disorder in which people did not take in enough air; they did not breathe enough. It was also widely believed that the reason people with this syndrome were sleepy during the day was that their carbon dioxide levels were too high when they were awake. But when we tested the patient's carbon dioxide level, it was normal, as was the result of every other blood test for abnormal states of consciousness. Her results proved that she did not suffer from Pickwickian syndrome. Furthermore, her symptoms did not fit any other syndrome based on what I knew or had read about. I was as stumped as the specialists at the hospital who had consulted on her case. I remember thinking that if I had been a bit smarter or if we had known more, I would have been able to help her.

Roughly two years later, while I was a medical resident at the Royal Victoria Hospital in Montreal, I had a case that was nearly identical to that of my patient in Chicago. This patient was a man, and he had one additional symptom—he had seizures while he slept.

One night, as I was doing my patient rounds, I observed that he stopped breathing when he slept, and I wondered whether this was somehow related to his sleepiness and his seizures. My speculations quickly led to what I believe to be the first sleep-breathing study in Canada and the publication of the first paper I wrote in the sleep field.

In the study we found that when he slept, the patient's breathing passage became repeatedly obstructed and he stopped breathing, and that when he stopped breathing, his heart rate dropped and at times his heart would stop for up to ten seconds. When the brain is deprived of blood because the heart has stopped pumping, seizures can occur. We now could explain both the patient's sleepiness and his seizures. We treated the patient with a tracheostomy, a hole placed in his windpipe, which bypassed the obstruction. His sleepiness was cured, and he never had another seizure.

I had found a case of sleep apnea, a disorder thought to be extremely rare, which had only been described a few years earlier, in the mid-1960s, in obscure (at least to me) European medical journals. This was probably the disorder that should have been diagnosed in the Chicago patient as well.

If you were to search the National Library of Medicine databases on the internet for all the articles written that contain the keywords *sleep apnea syndrome*, you would find more than six thousand articles; yet not one of them is listed before 1975.

But sleep apnea is not new; though it was not recognized clinically until the 1970s, it has been around for thousands of years, as can be seen in Dickens's Fat Boy and in historical figures from our earliest times. It has been around as long as there have been obese people.

In 360 B.C.E., the tyrant of Heraclea (now called Iraklion, on the island of Crete) was a man by the name of Dionysius, a contemporary of Alexander the Great. He was so overweight that during public appearances he allowed audiences to see only his head. Historical texts reveal that he had a tendency to fall into a very deep sleep, so he hired people to poke him with long, thin needles while he slept, presumably to keep him breathing. Apparently they did not do the trick; the same texts note that he eventually "choked" on his own fat.

The U.S. president William Howard Taft, who was elected in 1908, had sleep apnea while he was in office, although none of his doctors understood the cause of his problem. He was overweight, he snored, and he was sleepy throughout his four-year term. When he became professor of law at Yale in 1912 after leaving the White House, he was still obese, requiring extra-wide chairs. He soon lost much of his excess weight and apparently was cured of his apnea. Eight years later, he became chief justice of the U.S. Supreme Court, where he demonstrated none of the symptoms of apnea.

CAUSES OF SLEEP APNEA

Two types of problems lead to sleep apnea. The most common is an obstruction in the upper respiratory tract that can result in *obstructive sleep apnea*. Usually the breathing passage is kept open by muscles of the upper airway, but a variety of problems can impair the muscles' ability to keep the passage open. Air going into the lungs usually travels through the nose, then makes a turn and moves behind the soft palate and down the throat (pharynx), before it finally gets into the lungs. A condition that interferes with the flow of air to any of these locations can result in sleep apnea. Thus, anything from a blocked nose to enlarged tonsils to a narrowed breathing passage due to obesity can lead to obstructive sleep apnea. For people with this disorder, the breathing passage is open while they are awake but obstructs when they are asleep; they cannot sleep and breathe at the same time.

In children apnea could be caused by enlarged tonsils and adenoids, obesity, and even an abnormal (often inherited) jaw or facial structure. Just as we now see a great deal of type 2 (adult type) diabetes in children as a result of the obesity epidemic, we now see a great deal of apnea caused by obesity in children.

In adults apnea is most often caused by obesity or, as in children, by abnormal (often inherited) jaw or facial structure.

In the less common type, the problem is in the central nervous system. This results in *central sleep apnea*, a reduction in the electrical impulses from the nervous system to the muscles used for breathing. This disorder sometimes occurs when there are abnormalities in the nervous system and can also occur in people with heart failure, in people who have had a stroke, or as a reaction to narcotic pain medications.

When people stop breathing, the level of oxygen in the blood goes down and the level of carbon dioxide goes up. The low blood oxygen level forces the cardiovascular system to work harder, and the changes in the heart rate and the autonomic nervous system can increase the blood pressure. The increases in carbon dioxide level also affect the circulation, particularly the circulation of the brain. As a result, patients may awaken with headaches. For breathing to resume, the brain needs to wake up and open the breathing passage. People with sleep apnea awaken many hundreds of times per night. These disruptions result in a lack of quality sleep, which in turn causes severe daytime sleepiness.

In upper airway resistance syndrome (UARS), the variant of sleep apnea more common in women, breathing passages are not completely blocked; the awakenings are caused by snoring or snorts.

When we study people with sleep apnea, we see that the worst episodes occur while they are dreaming (in rapid eye movement sleep). One of the reasons for this is that people are paralyzed in REM sleep, and so the muscles that keep the airway open are also paralyzed. In addition, the body's defense mechanisms are inhibited in REM sleep. Normally, there are systems that protect us against low blood oxygen and high carbon dioxide levels. They make us breathe more deeply and will wake us up if necessary. These alarm systems seem to be suppressed in REM sleep; the defense mechanisms do not start to work until the oxygen level is very low and the carbon dioxide level very high. In some women with sleep apnea, the only time their breathing becomes abnormal is during REM sleep.

Recognizing Sleep Apnea

When I was a medical student, I was taught that if we could understand everything there is to know about syphilis, we would know all there is to know in medicine. Of course, this was an exaggeration, but the point was that syphilis is a disease that can affect many different organ systems, including the neurological system and the cardiovascular system. In addition, patients could have many different symptoms, so by understanding syphilis, a medical trainee could learn about all these systems and thus understand many aspects of internal medicine and microbiology.

I believe that the same thing is true of sleep apnea. If we knew everything about sleep apnea, we would know a great deal about medicine. Sleep apnea affects many organ systems. People with sleep apnea have a bewildering array of symptoms that take them to the doctor. Perhaps one of the reasons it took so long for medical science to recognize the disorder was its multitude of symptoms. Additionally, the symptoms in men and women are not always the same.

But another reason sleep apnea was missed was that sleepiness was never considered a symptom. It was not something doctors asked patients about; people with sleepiness were dismissed as either not getting enough sleep or lazy.

The most important cluster of symptoms of sleep apnea is sleepiness, snoring, and stopping breathing during sleep; for women, insomnia is also an important symptom.

SLEEPINESS

People with sleep apnea often fall asleep in low-stimulus situations such as watching television, waiting at a doctor's office, and traveling as a passenger in an automobile. Additionally, even small amounts of alcohol can dramatically worsen their sleepiness. They might fall asleep at times when it is dangerous to do so, such as when they are operating a motor vehicle or piloting an aircraft. The two most unusual circumstances in which patients of mine with apnea fell asleep were during their own wedding ceremony (the groom started snoring while still standing up) and during sexual intercourse. In both instances, the spouses insisted on immediate consultation!

SNORING (AGAIN)

Many people with obstructive sleep apnea believe they sleep well. Because they cannot hear their own snoring, they do not think that they do snore or that their snoring is disruptive to others. In our sleep clinic, we have patients look at a digital video of themselves sleeping that is taken during the sleep test. They often remark, "*My God, is that me? What have I put my family through?*"

Alcohol makes snoring louder and more severe. Someone who normally has no other symptoms of apnea than snoring might stop breathing while sleeping after drinking even small amounts of alcohol—alcohol can turn snoring into apnea.

STOPPING BREATHING

The word *apnea* means "stopped breathing." Though bed partners might be tempted to think that any silence emanating from a snorer would be a good thing, nothing could be farther from the truth. When a person who snores suddenly becomes silent, the listener usually waits with trepidation for the breathing (and the snoring) to resume. This can happen over and over again during the night. What is worse to the listener than the loud snoring is the hundreds of repetitive cycles of noise, quiet, noise, quiet. These apneic events make up the third of the main features of sleep apnea. However, there are other symptoms.

OTHER SYMPTOMS

Stopping breathing affects several organs of the body, which can cause many other symptoms and be even more of a problem than the three main symptoms discussed so far. People with sleep apnea may complain of awakening with choking or headaches (either during the night or in the morning), loss of interest in sex, needing to take frequent trips to the bathroom at night, symptoms of cardiovascular disease, and heartburn.

In some patients, the clinical findings mimic a psychiatric disorder; people might complain of symptoms that are similar to those found in depression or other conditions. Women, in particular, are frequently treated for depression before their apnea is recognized.

OBESITY

The world is in the middle of a major epidemic of obesity that is affecting all age groups. It is estimated that by 2025, 18 percent of all men and 21 percent of all women worldwide will be obese. Currently, about two-thirds of American adults (compared to less than one-quarter forty years ago) are above their ideal weight, which is measured by the body mass index (BMI) and is defined as a BMI of less than 25. (The Centers for Disease Control offers a BMI calculator at its website, cdc.gov.) In 2016 the CDC reported that 35 percent of American men and 40 percent of women are obese (have a BMI of more than 30). The proportion of North Americans with extreme obesity increased more than six-fold between 1960 and 2012. Half of all people with extreme obesity are likely to have sleep apnea. They are also much more likely to have cardiovascular disease and diabetes. About 75 percent of sleep apnea patients at most sleep clinics are obese, and symptoms often started after a substantial weight gain. The average adult sleep apnea patient in our clinic has a BMI of 33. Even children are becoming obese, and many of them are now also suffering from sleep apnea.

Young, apparently healthy athletes who are overweight are subject to sleep apnea. For example, a study in 2016 estimated that 8 percent of college football players have sleep apnea, and a study in 2003 reported that about one-third of NFL linebackers have it. Another study in 2010 reported that about half of retired NFL players had sleep apnea.

RECOGNIZING SLEEP APNEA IN OTHERS

From the symptoms mentioned in this chapter, it should be relatively easy to recognize when a sleep-breathing problem is present. Listening to the loud, struggling snoring sounds—and the quiet periods when breathing is obstructed—is frightening. But often the symptoms come on so slowly that their significance is missed until something dramatic happens, an event that becomes a wakeup call to the family. This could include a patient's falling asleep at the wheel, missing important appointments, or nodding off at unfortunate times. In general, if a bed partner snores, is observed to stop breathing while asleep, has had changes in personality (irritability, inability to concentrate, and so on), or falls asleep at the wrong time and in the wrong place, he or she may well have sleep apnea.

Many people, even those with severe apnea, might doubt that they have a problem and think that they are champion sleepers. Only when they see and hear themselves sleeping does the significance of their problem become apparent to them. The spouse or bed partner should persevere to make sure that the problem is evaluated—perhaps by making a video of the person sleeping.

As noted, sleep apnea can occur in people of any age, including children. The symptoms and causes of sleep apnea in children might be different from those of adults, and parents should keep a few general rules in mind. If a child snores loudly most nights and is observed to stop breathing, it may be a signal that he or she has apnea. If this symptom is associated with very restless sleep such as moving the neck or the jaw to open the breathing passage, that is another indicator. A child who has large tonsils, who is overweight, or who has a small jaw may have an obstructed breathing passage. Orthodontic evaluation and treatment can often cure the apnea if it is caused by abnormalities of the jaw.

Children who are sleepy may appear to have attention deficit hyperactivity disorder; in other words, they may seem hyperactive rather than sleepy. Sleep apnea is so common in children that in 2012 the American Academy of Pediatrics suggested that all children be screened for sleep apnea during routine office visits.

HOW MANY RISK FACTORS DO YOU HAVE?

The Berlin Questionnaire shown in the figure below assesses a person's risk of having sleep apnea. I have modified the questionnaire slightly to include not just overweight people but also those who might have other abnormalities such as small jaws that could lead to a sleep-breathing disorder. Like any tool that estimates risk, it could overestimate or underestimate. If you think you have a sleep-breathing problem, you should describe the symptoms to your doctor as accurately as possible, perhaps bringing a family member with you. Take a copy of this completed questionnaire when you go to the doctor to demonstrate your reasons for concern.

The questionnaire indicates the statistical likelihood that a person has apnea. Like most tools, it is far from perfect. Some people who score positive might not have apnea, while some who score negative might have it. The questionnaire has been weighted to be sensitive, so that it will miss the fewest number of people who might have sleep apnea. Using this tool and the other tests described above, your doctor will be able to determine whether you should see a sleep specialist.

Our Untreated Sleep Apnea Sufferers

As is true of most people who have sleep disorders, the average person who is diagnosed with sleep apnea will have seen many doctors before that correct diagnosis is made. One study showed that sleep apnea patients were seeing doctors more frequently than normal for as long as ten years before they were correctly diagnosed. Most doctors do not question their patients about

Category 1 SNORING
❑ Do you snore?
❑ Is snoring louder than talking?
❑ Is snoring present at least 3-4 times a week?
❑ Has snoring ever bothered others?
❑ Has anyone noticed the stopping of breathing during sleep at least 3-4 times a week?
___Add up the number of positive answers
✓ If more than 2 this category is positive.

Category 2 SLEEPINESS
❑ Are you tired or fatigued after waking up more than 3-4 times a week?
❑ During the day are you tired or fatigued more than 3-4 times a week?
❑ Do you have any trouble staying awake driving?
___Add up the number of positive answers
✓ If more than 2 this category is positive.

Category 3 RISK FACTORS
❑ Do you have high blood pressure?
❑ Is your BMI more than 30 or is your neck collar size more than 17 inches?
❑ Do you have a very small jaw or a large overbite?
___Add up the number of positive answers
✓ If more than 2 this category is positive.

FILL THIS IN LAST: Check the positive categories.
❑ Category 1 SNORING
❑ Category 2 SLEEPINESS
❑ Category 3 RISK FACTORS
___Total the number of positive categories
✓ If more than 2 boxes are checked the chance that apnea is present is high.

The Berlin Questionnaire

how they sleep, whether they snore, or whether they feel sleepy in the daytime. Many doctors still believe the stereotype of apnea as a disease of obese middle-aged men; they often discount the symptoms of apnea in the women and children they see.

The best data available, based on studies done by the University of Wisconsin first reported in 1993 and continuing to this day, indicate that sleep apnea occurs in at least 2 percent of women and 4 percent of men in the United States; similar data suggest that these percentages also obtain throughout the Western world. This means that if a family doctor sees one hundred adult patients, fifty women and fifty men, each week, three of them (two men and one woman) will have sleep apnea. Imagine how many patients a single doctor is likely to miss in a year!

In our sleep clinic, the average age of patients being diagnosed with sleep apnea is around fifty, but many have had symptoms for five to ten years before they are diagnosed. Some of our patients have lost their jobs and homes because of the disorder. Many are being treated for conditions that they may not have, such as depression, and receiving unnecessary medications that could have serious side effects. Patients with sleep apnea are more likely to develop high blood pressure or suffer a heart attack, heart failure, or a stroke.

Many of my patients have fallen asleep driving—and some of those patients have been truck drivers, train engineers, and airplane pilots. Studies in several countries have shown that apnea patients are at much greater risk of having car accidents. Doctors need to make the connection that patients who fall asleep while driving could have a major sleep problem. In December 2013 a Metro-North Railroad engineer with undiagnosed sleep apnea fell asleep as he was driving a train in the Bronx, New York, resulting in a derailment. Four people were killed and sixty-one passengers were injured. As a result of this and other high-profile accidents, the public is increasingly demanding that locomotive engineers be tested for sleep apnea. Some trucking companies already screen and test their drivers for sleep apnea.

Management of Sleep Apnea

If your doctor suspects you have sleep apnea, he or she will ask about sleep and sleepiness during the daytime. If the doctor strongly suspects you have sleep apnea, he or she will probably refer you to a sleep clinic for overnight testing. This testing can also be done in the patient's home, but the home testing is not usually as comprehensive as the testing done in a laboratory.

WHAT THE SLEEP TEST SHOWS IN SLEEP APNEA

To prove that apnea is present, the sleep test evaluates brain waves (via an EEG) to determine whether the person is sleeping; eye movements (by means of electrodes around the eyes) to see when the patient is in REM sleep; the heart rhythm (with an electrocardiogram); blood oxygen levels; measures of effort to breathe by the chest and abdomen; and whether the person is breathing from an airflow indicator in front of the nose and mouth. These measurements are monitored over an entire night.

The figure below shows a few of the measurements made in a sleep study. These measurements show that the patient has sleep apnea. The section on the left illustrates what we see when the person is awake. The section on the right shows what we see when the person is asleep. When the person is awake,

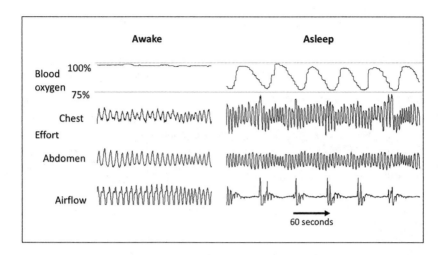

Measurements Made in a Sleep Study

breathing effort and airflow are regular and blood oxygen level is steady. When the person is asleep, although efforts to breathe continue, there are times when airflow is zero. The blood oxygen level drops to dangerously low levels with each episode of stopped breathing and the oxygen level is no longer steady; it now goes up and down with each episode. The episodes in this example occur about once per minute.

Not included in the example is the EEG, which showed an awakening right before the patient started breathing again with each episode. The typical patient in our clinic stops breathing and wakes up about thirty to forty-five times every hour. Stopping breathing less than fifteen times an hour indicates mild apnea; between fifteen and thirty times is moderate; more than thirty times is severe. We also digitally video record the entire night and show patients their record and video the next morning. When they examine the results of their sleep study, they often become frightened and more aware of the dangers associated with these events, and realize that they must get treatment.

WEIGHT LOSS AND GENERAL MEASURES

Treatment of sleep apnea includes the measures used for all people who snore (weight loss, avoidance of alcohol and certain drugs, use of dental appliances), described earlier in this chapter, along with specific treatments to open the airway that becomes blocked during sleep.

When I diagnose patients with sleep apnea, I also point out to them something that I have observed many times. The forty-five-year-old sleep apnea sufferer is still mobile and might be in reasonable health. The fifty-five-year old sleep apnea sufferer who is overweight will probably have already had a major cardiovascular event such as a heart attack and frequently has developed diabetes. Arthritis in the knees or the hips may also be present. The hips and knees are particularly vulnerable because those joints are under the greatest amount of stress when a person is overweight. With knee and hip problems, the patient is no longer as mobile as he or she once was, and so exercises less, causing the weight to balloon upward. I often hear patients use the excuse, "I can't lose weight because I can't run or walk on the treadmill or use the bicycle. There is no exercise I can do." I tell them to go to their local swimming pool. Exercise programs are available or can be designed for these patients. Walking in the water, which removes pressure from weight-bearing joints, is excellent exercise. Other, supervised exercises in water can also be beneficial. Along with regular exercise, the patient needs to normalize his or her food intake. It is beyond the scope of this book to focus on dieting, but if a person cannot lose weight, there may be another medical condition underlying his or her obesity; such patients should seek professional help. In cases of extreme obesity bariatric surgery might be an option.

OPENING THE BLOCKED BREATHING PASSAGE

When obstructive sleep apnea has been confirmed by testing, the treatments mentioned earlier for snoring are usually recommended. In cases that don't respond to these treatments, if severe obesity or apnea is present, or for very severe cases we recommend more aggressive treatment, especially continuous positive airway pressure (CPAP).

CPAP. With this treatment, the patient wears a mask, usually over the nose, but sometimes over the nose and mouth. This is attached by a hose to a device the size of a toaster which generates pressure that opens the breathing passage. CPAP treatment usually gets rid of snoring, effectively reestablishing regular breathing. This type of treatment does not work for everyone because some patients have difficulty getting used to the pressure in the nose or develop symptoms of a blocked or runny nose. Sometimes, humidification is added to the system, but some people still have difficulty. CPAP seems to be tolerated and effective in about 70 percent of severe sleep apnea cases. For bed partners concerned that the noise of the machine is going to replace the noise of the snoring patient, with no net gain in quietness, the machines widely available in 2017 are fairly quiet—rather like a very quiet air conditioner.

Other PAPs. The CPAP machine creates a single effective pressure. Machines are also available that continually adjust themselves (these are called autotitrating or AutoPAP), as well as machines that deliver two pressures (called Bilevel or BiPAP). Some machines help the patient breathe in addition to opening up the breathing passage. The sleep specialist will generally recommend the machine type and settings, as well as the mask type. We can now monitor remotely how well the patient is doing on treatment by communicating with the machine wirelessly. Patients can also monitor how they are doing by using a smartphone that communicates with their PAP device.

Dental appliances. In addition to helping with snoring, custom-fitted (by a dentist or orthodontist) oral appliances can be very effective for some people with apnea. The appliances are worn at night only, and they bring the lower jaw up and forward. Since the tongue is attached to the lower jaw, moving the jaw forward moves the tongue forward, thus increasing the size of the breathing passage behind the tongue.

Provent. This treatment was introduced about 2010. It consists of a miniature disposable one-way valve inserted into the nostrils and attached to the nose by an adhesive strip.

Surgery. If apnea is caused by an obvious problem, such as enlarged tonsils or an abnormal structure of the jaw, surgery might cure the problem. In very severe cases that do not respond to CPAP, the patient might require a tracheostomy, as my Canadian patient did. The patient breathes in and out of a hole cut in the front of the neck rather than the mouth or nose. Until the mid-1980s, this was the only treatment available, and it was usually effective, although highly invasive. This is the treatment of last resort for people severely affected with sleep apnea.

Another type of operation that has been performed for snoring and apnea is to remove tissue from the back of the soft palate including the uvula, the tissue that hangs down at the back of the throat. This type of surgery has been done using scalpels, lasers, radio waves, and most recently robots. As mentioned earlier, many people do not respond well to such surgeries; thus most sleep specialists (myself included) do not recommend surgery as the first treatment.

In 2014 a new surgical treatment that involves stimulation of a nerve that goes to the tongue was approved in the United States by the FDA. The stimulation is achieved by the same type of device that is used in heart pacers. It is implanted under general anesthesia by an ear, nose, and throat surgeon. This is only recommended after other treatments have been exhausted and is not recommended if the patient is too obese (has a BMI of greater than 30).

TREATING SLEEPINESS

Because people with untreated sleep apnea stop breathing and then awaken repeatedly when they sleep, they do not find naps refreshing; they usually wake up feeling groggy. Some people find that they sleep best when resting upright, and these people may benefit from naps.

The treatment for the sleepiness of sleep apnea is to open the blocked breathing passage. A patient with apnea who is on CPAP should use the CPAP every time he or she sleeps, including naps. The FDA has approved the use of the wake-promoting drug modafinil (known as Provigil, or Alertec in Canada) and armodafinil (Nuvigil) to treat patients who are still sleepy even after using a CPAP device.

OPERATING MOTOR VEHICLES

In many parts of the world, a person diagnosed with sleep apnea will not be allowed to operate a motor vehicle until she or he has been treated. In most places, driving a car is considered not a right but a privilege—and that privilege can be withdrawn or suspended. This applies not only to people who have sleep disorders but to those with any medical problem that might endanger themselves or the public. This would include people with epilepsy and those who have had a recent heart attack or stroke. Since the regulations vary from place to place, people with sleep apnea should check with their doctor about the regulations in their state.

In many places, commercial drivers (for example, those who operate buses or tractor trailers) are screened by questionnaire, BMI, and neck size. If they are deemed to be at high risk, they will be given a sleep test. If they are found to have sleep apnea, they must be treated and be able to prove that they are using the treatment.

I am reminded of a patient who finally sought help, but not until he fell asleep driving and crashed his station wagon with three of his grandchildren in the backseat. Luckily, they all survived. Sometimes the symptoms come on so slowly—over a period of years—that the person is not aware of them and believes that he or she has a normal sleep-wake pattern.

I was once invited to give a lecture to a group of family physicians about sleep disorders. The room was small, and the doctor who invited me to speak, a thin young man, promptly fell asleep when my presentation began. I awakened him to demonstrate to the rest of the audience how to do an interview with a patient who might have a sleep disorder. When I asked him whether he fell asleep while driving, he said, "Doesn't everybody?" This doctor had been falling asleep while driving his entire adult life and believed that this was a normal state of affairs. He had a classic history of obstructive sleep apnea caused by a small jaw, and his symptoms disappeared when he was fitted with an oral appliance, which he now wears at night.

SLEEP APNEA DURING PREGNANCY AND EARLY MOTHERHOOD

For some pregnant women apnea is present before the pregnancy begins, while in other women it can come on if there is a massive weight gain during pregnancy. Pregnant women with apnea should be evaluated and treated because they might be prone to high blood pressure or a more serious problem, preeclampsia (see Chapter 4). There is medical evidence that suggests that babies born to pregnant women with sleep apnea could be smaller than average. If the blood level of oxygen is too low in the mother when she sleeps, it will be too low for the baby. If the woman is not treated, the symptoms and the sleep quality will tend to worsen as the pregnancy progresses. If she is working and accommodations such as working a reduced number of days or a shortened workday are not possible, she may need to consider taking a medical leave. It is best for pregnant women who have untreated sleep apnea to not drive a car.

After the baby is born, the mother must be alert to take care of it. Mothers diagnosed with sleep apnea should thus be started on CPAP (or at least sleep sitting in a 45-degree position) as soon as the diagnosis is made and continue using it after the baby is born. Mothers with apnea should continue using the CPAP until they have lost weight and the apnea is cured.

MENOPAUSE

After the onset of menopause women are about three times more likely to develop sleep apnea. There are several reasons for this. The levels of the sex hormones that protect the women against apnea decrease during menopause; additionally, many women put on weight at this time. But not all menopausal women with sleep apnea are obese. Research reported in mid-2003 showed that hormone replacement therapy seemed to improve sleep apnea in some postmenopausal women. This is currently an area of active research.

Back to the Farmer's Daughter

When I examined the girl's throat with my trusty flashlight, I saw a giant set of tonsils, each one almost the size of a golf ball, meeting in the middle of her throat and virtually blocking her breathing passage. They were the cause of her apnea, and I referred her to an ear, nose, and throat surgeon. After removal of her tonsils and her adenoids, her breathing became normal. Her alertness returned, and she went back to school. I wish every medical problem had such a happy ending, but I also wish that my young patient had not had to suffer needlessly for so long.

What happens during sleep, which makes up roughly a third of a person's life, has enormous importance. A good night's sleep can energize a person for daytime activities, but a bad night's sleep, or a bad sleep disorder, can put a person's life in danger. Both patients and doctors need to be attuned to their sleep so that the symptoms of sleep apnea do not go untreated.

13

Narcolepsy

THE MYSTERY. How is it possible for people to dream while awake? How can a sleep problem affect muscle tone? The devastating disorder narcolepsy is often not diagnosed until years, sometimes decades, after the symptoms start.

~~~~~~~~~~~~~~~~~~~

The Case of the First-Year Med Student

At the end of a lecture I gave to a first-year medical school class, seven or eight students rushed down to speak to me about their personal sleep problems. I was not surprised: medical students often think they have whatever illness they've just learned about. All but one had a problem with daytime sleepiness caused by not getting enough sleep because of their demanding schedule. After listening to lectures most of the day and studying long hours each night, most students come to class toting a large cup of coffee, and several nod off during the class. One sleepy young student had another symptom that I had mentioned during the lecture. Unfortunately, she had ignored this symptom for several years, assuming—incorrectly—that everyone had it.

She told me that about twice a week she would awaken from sleep in the middle of a dream and discover that she was absolutely paralyzed. She could breathe, but she could not move her arms, legs, or head, nor could she speak. The problem had started about five or six years earlier, and it had terrified her, especially when she dreamt that a devil-like creature was staring at her. The paralysis was not her only symptom. She also experienced an onrush of dream imagery as she was falling asleep and at times even before she started to fall asleep. Sometimes she had a vivid dream but could not tell whether she was awake or asleep. Because she had experienced these symptoms for so long, and because her sister had identical symptoms, she assumed that everyone had them. But everyone doesn't.

What Is Narcolepsy, and Who Gets It?

Narcolepsy is a chronic neurological disorder caused by abnormal brain chemistry, which leads to a perplexing constellation of symptoms that may include one or more of the following: severe sleepiness, vivid dream imagery upon falling asleep (known as hypnagogic hallucinations) or waking up (hypnopompic hallucinations), waking up paralyzed (sleep paralysis), and sudden-onset temporary muscle weakness (cataplexy). The most common symptom is falling asleep at inappropriate times and places. Narcoleptic patients frequently have difficulty falling asleep and sleeping through the night. Narcolepsy most often comes on in the mid-teenage years, and it affects women and men equally.

Scientists around the world have been trying to determine how common narcolepsy is, and the answer seems to vary from country to country. Narcolepsy appears to be very common in Japan, whereas in North America it is estimated that it affects approximately 1 out of every 2,000 people. Based on research studies, it is estimated that in the United States there are between 100,000 and 150,000 people with the disorder. Most of them have not been diagnosed.

One study of a large group of patients (63 percent women, 37 percent men) in 1997 reported that most had had symptoms for roughly fifteen years before their narcolepsy was correctly diagnosed. People with narcolepsy are often treated for some other condition, often depression, which they might not have. Children might be misdiagnosed with attention deficit disorder or attention deficit hyperactivity disorder (ADD or ADHD) when they are simply too tired to pay attention. Some doctors assume that a woman who is sleepy probably has depression. Not only does the incorrect diagnosis delay the correct treatment, some of the drugs used to treat these disorders can make the narcoleptic's sleepiness worse. Because the disorder comes on often during the middle teenage years, people with narcolepsy are left untreated at precisely the time when they are going to school and developing the skills that will carry them through the rest of their lives. Teenagers with undiagnosed narcolepsy might feel that they cannot cope with school and drop out. Narcoleptic patients are also at much greater risk of having automobile accidents. Narcolepsy can be especially devastating for caregivers who also work outside the home. Keeping a job is difficult when a person cannot stay awake. Maintaining a relationship is a challenge. Raising children and keeping up with their busy schedules is even more exhausting. Simply put, narcolepsy can ruin lives. It is essential that sufferers get a correct diagnosis as soon as possible.

Narcolepsy seems to be due to the brain's abnormal regulation of rapid eye movement sleep. As we have seen, during REM sleep (dreaming sleep), humans are paralyzed. (This phenomenon has also been observed in virtually all higher life forms including other mammals and birds.) During REM sleep, the muscles of the body are paralyzed except the ones necessary to sustain life, such as the breathing muscles (diaphragm), the heart and other muscles that are controlled within themselves, and some of the muscles at the top and bottom of the gastrointestinal system.

Scientists have made great strides in the understanding of the brain circuits and chemicals that are involved in REM sleep. But they still cannot explain why we dream and why we are paralyzed during REM sleep, though one hypothesis is that our bodies are paralyzed so that we cannot physically react to our dreams. Normally, adults do not experience the first episode of REM sleep until they have been asleep for about ninety minutes. Thereafter, they will have an episode of REM sleep at roughly ninety-minute intervals. Most people will dream three to five times a night, and while they are dreaming, they are paralyzed. This is normal sleep.

Patients with narcolepsy fall into REM sleep at the wrong time and in the wrong place. The dreamlike imagery that they see at the onset of sleep or even while they are still awake is called a hypnagogic hallucination. Sometimes when a patient with narcolepsy awakens from a dream, the paralysis of REM persists. One eighteen-year-old narcoleptic described his dreams: "I have seen a man at the window with a pointed face, and when he sees that I have seen him he turns around to leave and the entire back of his head is a metal plate that is actually screwed into his head. This is one dream that I had when I was very young and still have recurring dreams with him in them. At the time of this dream I had difficulty believing that this man was not real because I thought I was awake."

Although we do not know for certain what causes narcolepsy, many clues are coming in from unexpected sources, such as research on Doberman pinschers and mice. (Narcolepsy has been found in several breeds of dogs, including not only Doberman pinschers but Labrador retrievers, poodles, dachshunds, and some mixed breeds.) Science made a huge leap forward recently when a new chemical culprit was discovered that might play a major role in causing narcolepsy. This chemical is known by two names, orexin and hypocretin. Research suggests that patients with narcolepsy do not produce sufficient quantities of this chemical, or that receptors in the nervous system might have stopped responding to it adequately. Experiments have been done in which the genes responsible for the production of this chemical are knocked out in experimental animals; the animals developed the features of narcolepsy, sleep attacks, and cataplexy. The discovery of this chemical and its function will help scientists and doctors better understand and treat this debilitating sleep disorder.

It is believed that narcolepsy has both a genetic component and an autoimmune component. People are not born with the disorder. The symptoms usually appear unexpectedly, although they sometimes show up after a mild infection, a traumatic brain injury, or a concussion. Specific gene variations have been reported recently that show a strong association with the age when some of the symptoms appear.

Narcolepsy can apparently also be triggered by chemicals. In the winter of 2009–10 the world was trying to contain an epidemic of the H1N1 virus (swine flu). A vaccine containing an "enhancer," a chemical called ASO3, was widely used in Finland, Sweden, and the United Kingdom. Some children and young adults with genetic susceptibility who were vaccinated with this product developed narcolepsy. (This vaccine was never approved for use in the United States.)

One other symptom of narcolepsy, which can be very upsetting to the sufferer, is the experience of a sudden loss of muscle control. This sometimes happens when patients are awake and they become excited. After hearing a joke, for example, a narcoleptic person might feel some of the manifestations of REM sleep come on. These lead to a form of temporary paralysis called cataplexy that can cause the individual to collapse in a heap. One patient said that she felt like a puppet with all the strings cut off whenever she heard a joke. Yet the person is awake and conscious, even though he or she cannot move. Women with cataplexy might not have orgasms during sexual intercourse because they try to avoid arousal that might lead to a cataplexy attack. At other times, the loss of muscle tone is more limited and might involve only the muscles of the face or the neck. These episodes are sometimes misinterpreted as an epileptic seizure.

Depending on the symptoms and with recent understanding of the brain mechanisms, doctors have defined two types of narcolepsy: narcolepsy with cataplexy (now called type 1 narcolepsy) and narcolepsy without cataplexy (now called type 2 narcolepsy). Scientists believe that type 1 narcolepsy is caused by low levels of the chemical orexin or hypocretin, and one way to test for the disorder is to measure the level of this chemical in the fluid that bathes the nervous system.

RECOGNIZING NARCOLEPSY IN OTHERS

In addition to recognizing their own symptoms of narcolepsy, parents and caregivers should be alert to symptoms in their children.

Young children. Recognizing narcolepsy in young children is difficult, but there are clues parents can watch out for. A child over the age of five who starts to take naps again might have narcolepsy. Children who fall asleep at the wrong time and in the wrong place—for example, at school, when watching television, or in the car—probably have an abnormal sleep condition. Teachers might tell parents that their child is very sleepy or daydreaming in the classroom. Some children complain of frightening nightmares, which could be hypnagogic hallucinations; parents should pay attention to how frightened the child seems to be and whether specific nightmares recur. One mother described her daughter's experience: "When she was much younger, she used to come into our bedroom afraid after a dream; we would make her a bed beside ours to sleep there the rest of the night. One particular dream was of a pointed-face man looking into her bedroom window. She was so scared that we actually investigated the window and outside to make sure someone hadn't

been there. We found that it was not physically possible for anyone to be looking in. This was still not enough for her, and we never completely convinced
her it wasn't real. She had many more dreams that unsettled her over the years
and some she just could not bring herself to relate to us."

The child's sleepiness might appear or be diagnosed as an attention deficit
hyperactivity disorder. Children diagnosed with ADD and ADHD are often
treated with methylphenidate (Ritalin), a central nervous system stimulant.
Thus children with narcolepsy misdiagnosed with ADHD or ADD will have
improvement of their sleepiness. Parents should be alert to all the other symptoms, and if they suspect narcolepsy, take the child to a sleep clinic.

Adolescents and teenagers. As we have seen, the onset of narcolepsy frequently comes during the teenage years. The teenager may start to sleep in
and have to be dragged out of bed, even though he or she has gone to sleep
at a normal time. He or she might fall asleep in school and experience a drop
in school performance and grades. People with narcolepsy are almost always
sleepy. They differ from normal sleep-deprived teenagers, who become alert
as soon as they have a few good nights of sleep. They also differ from teenagers
with circadian clock problems (see Chapter 8), who might fall asleep late and
wake up late, but who have no trouble staying awake in the afternoon.

The symptoms of these young patients are often interpreted as depression, and their narcolepsy goes undiagnosed and therefore untreated for years.
Sometimes children might become so discouraged by the constellation of
problems facing them that they start to avoid school. This can also be a warning sign for parents. When children do not want to go to school because they
fall asleep in class and their friends make fun of them, it is not a sign of laziness. At this time in an adolescent's life, when so many hormonal and growth
changes are occurring, the addition of the symptoms of narcolepsy can be
extremely difficult to deal with. It is essential that teenagers with narcolepsy
be diagnosed correctly as soon as possible.

DIAGNOSIS

As my colleagues and I reported in a medical article in 2002, family doctors
diagnose only about 20 percent of narcolepsy cases even when patients have
classic symptoms. One reason why narcolepsy is underdiagnosed is that most
doctors do not question patients about how they sleep. And many doctors

know little if anything about narcolepsy. In most medical schools, students receive just two to four hours of sleep medicine training during their entire educational process. Most of the patients with narcolepsy that I have seen have a classical clinical history.

For teenagers, often the first symptoms are a drop in grades at school. As a member of the Yale University faculty, I have seen a number of students for whom this was the first symptom. The students have difficulty staying awake in class, trouble concentrating, and difficulty completing their work. Sometimes parents blame their schedules, sometimes they assume the problem is depression. Doctors might even start the student on antidepressants, which could make matters worse.

Yet narcolepsy can be diagnosed in a few minutes if the doctor asks three or four specific questions.

Clinical interview and examination. When was the last time a doctor asked you whether you dream while falling asleep? This and a few other questions can help the doctor make a diagnosis of narcolepsy. The questions might include: Do you fall asleep at the wrong time and place? Do your knees buckle or feel weak if you hear a joke or become angry? Do you sometimes wake up and find that you cannot move? Do you dream during naps?

Because narcolepsy is a lifelong illness that will require lifelong treatment and because narcolepsy patients sometimes suffer from other sleep disorders as well, most sleep specialists will order a sleep test to confirm the diagnosis.

Sleep test. At the sleep lab, doctors perform two types of sleep studies, a nighttime and a daytime study.

The nighttime study (called a polysomnogram) can show the early onset of REM when the patient falls asleep, and the patient's sleep might be disrupted with many awakenings. Ironically, people with narcolepsy, who fall asleep too easily during the day, often have trouble falling or staying asleep at night. The personnel conducting the study should also look for other disorders that might cause sleepiness such as sleep apnea.

In the daytime study, called a multiple sleep latency test (MSLT), the patient is given four or five twenty-minute opportunities to fall asleep every two hours. If it takes the patient eight minutes or less to fall asleep on average, sleep scientists or technologists can diagnose severe sleepiness; if the person has REM sleep during two or more of the naps, it supports a diagnosis of narcolepsy.

IDIOPATHIC HYPERSOMNIA

Some patients experience profound sleepiness, but the clinical history does not include the REM-related symptoms that are the hallmarks of narcolepsy (cataplexy, hallucinations, sleep paralysis). When these patients are tested by MSLT, it confirms that they have a form of pathological sleepiness in spite of their sleeping eight hours or more a night. But if they do not have REM sleep during the opportunities to nap the test does not document narcolepsy.

In most cases the cause of the sleep disorder is unknown (hence the name *idiopathic*, which means "unknown cause"). Some patients may have had a concussion or traumatic brain injury. In most cases we do not know what brought on the condition.

Treating Narcolepsy

Medical science cannot cure narcolepsy at this time. Instead, doctors must treat the symptoms, recognizing that this is a disease that is not going to resolve on its own. (This is also true of people with idiopathic hypersomnia, for which doctors treat the symptoms—with, in fact, the same medications.) Though the individuals will probably have to use medication for the rest of their lives, their lives can be dramatically improved. They will be able to live a fairly normal life on treatment, and can be successful in their careers and other endeavors.

MEDICATIONS

Sleepiness, which is the most common and debilitating symptom of people with narcolepsy, can be treated with medications that make the patient more alert and help prevent the irresistible urge to fall asleep. There have been exciting developments in new medications, and research is ongoing.

Wake-promoting medications. Armodafinil (Nuvigil) and modafinil (Alertec in Canada, Provigil in the United States, the United Kingdom, and Australia), two related compounds, are the most commonly prescribed treatments to promote wakefulness. Both work on the specific parts of the brain that help maintain alertness. In contrast to stimulant medications, they have little effect on the body's other functions. These compounds might reduce the levels of estrogen (even those in birth control pills) in the blood, so females are warned to stop taking them when they are trying to become pregnant, and to use an additional form of birth control when they are trying not to become pregnant. Since patients with narcolepsy experience different degrees of daytime sleepiness in response to medications, the dosage needs to be customized to the individual. At this time, these two compounds, along with sodium oxybate, are the only medications approved in the United States for excessive daytime sleepiness associated with narcolepsy.

A reformulation in the past two decades of an old medication (gamma hydroxybutyrate) has been approved for narcolepsy under the name sodium oxybate (Xyrem). This medication, when taken at night, reduces cataplexy the following day. In addition, the medication improves daytime alertness. It is very short acting and thus is usually taken at bedtime and again four hours later. A newer formulation is being developed; this will allow the patient to take a single dose at night. There are tight controls on the use of sodium oxybate. Some antidepressants have also been given at bedtime to reduce REM-related symptoms.

Methylphenidate (Ritalin). Methylphenidate is widely known as the medication prescribed to children with attention deficit hyperactivity disorder. Paradoxically, the drug, used to wake up people with narcolepsy, is also used to calm down people with ADHD; it allows them to focus on their tasks. Besides having an effect on the central nervous system that makes individuals more alert, this medication also affects the sympathetic nervous system, which controls how some of our organ systems work. Stimulation of this part of the nervous system can result in an increased heart rate, rise in blood pressure, and a jittery feeling in some people. Although these symptoms may decrease with time, there is some concern that long-term use by people with narcolepsy could lead to adverse effects on the cardiovascular system. Thus, most sleep experts no longer consider methylphenidate the first drug of choice for increasing alertness in narcolepsy patients.

Amphetamines. For many people, amphetamines are "speed," illegal street drugs. But amphetamines have been prescribed by doctors for decades for a variety of conditions, and they have been used to treat narcolepsy since the 1930s. Many patients are still treated with them. Amphetamine molecules exist in two forms, which are chemically mirror images of each other. The brand names of preparations in the United States include Adderall (made up of both forms), Dexedrine (made up of one of the forms), and Vyvanse (a chemical that the body turns into one of the forms).

Amphetamines have a powerful stimulant effect on the brain and the sympathetic nervous system. They can also cause increases in heart rate and blood pressure, and they may cause sweating and jitteriness as well. Although amphetamines present the potential for abuse, I have seen this only rarely in my clinical practice. But the potential for abuse has led many countries (though not the United States and Canada) to remove them from the market, which can make crossing borders with them difficult. In some places, the prescription regulations for amphetamines (and other drugs that tend to be abused) are so strict that doctors do not prescribe them. (In some countries, doctors are required to fill out a triple-copy prescription form: one copy of the prescription stays in the doctor's file, one goes to the pharmacy, and a third goes to the medical licensing authorities, who monitor the usage of such medications. Prescriptions cannot be refilled over the telephone, and only short-term prescriptions are allowed.) In spite of these difficulties, some doctors—perhaps because they are not familiar with modafinil, described above—still prescribe amphetamines as soon as narcolepsy is diagnosed. I rarely prescribe an amphetamine preparation as the first treatment for narcolepsy.

NAPPING AND SCHEDULE ADJUSTMENT

Imagine patients' surprise when I prescribe naps. Naps can be extremely therapeutic for people with narcolepsy. A short nap of fifteen to thirty minutes can sometimes result in several hours of markedly improved alertness. (Long naps can leave a person feeling drugged or dopey and are not recommended.) Often a nap around lunchtime is sufficient to keep narcoleptic patients alert for several hours; some people may require a second nap later in the afternoon around 4:00 or 5:00. Children with narcolepsy can benefit from a short nap during lunch period at school.

An important aspect in the life of a patient with narcolepsy is finding ways to adjust his or her daytime schedule. Patients might need to take a nap during the day. Schoolchildren might need the school nurse to administer medications or to be allowed extra time to complete an examination. Patients seeking such accommodations find that a letter from the doctor can be helpful.

Living with Narcolepsy

DEALING WITH OTHERS

Hollywood has not been kind to people with narcolepsy. In the movie *Deuce Bigalow: Male Gigolo* (1999) one of the characters is a woman with narcolepsy who is shown falling face first into a bowl of soup. In *Bandits* (2001) a bank manager collapses from the excitement of being taken hostage and forced to open a vault.

Whether it is because of the jokes about narcolepsy in popular culture or because the condition is so misunderstood and stereotyped, people with narcolepsy often feel ashamed of their condition or worry that they won't be taken seriously by co-workers, supervisors, and others. An important aspect of living with narcolepsy is explaining the disease to family and friends. They need to be very direct: "No, I am not stupid. No, I am not lazy. No, I am not bored by what you are saying." Parents of children with narcolepsy should notify school administrators and teachers and others with whom the child interacts about their child's problem and its symptoms, explaining that if the child falls asleep it is not a sign of lack of respect or laziness but the result of a neurological disorder. In my experience, when someone with narcolepsy educates the members of his or her community about the disease, people will try to make proper accommodations. For one of my child patients, for example, the school set up a room where she could nap after lunch, which resulted in improvement in her academic performance.

DRIVING

A crucial aspect of daily life that is affected by narcolepsy is driving a car. Driving regulations vary substantially from place to place; in some states people with narcolepsy cannot drive unless they are undergoing treatment. When a child with narcolepsy is about fourteen or fifteen years old, parents need to have a conversation with him or her about the implications of the disorder as it relates to operating a motor vehicle. Even while on treatment, the narcoleptic must be taught about the importance of napping and not driving late at night when the medications might have worn off.

Government regulations in most developed countries do not allow people with narcolepsy to have commercial motor vehicle licenses: for example, they may not drive tractor trailers or buses.

PREGNANCY AND PARENTING

Because narcolepsy usually comes on during the teenage years, it is likely that women with the disease will have to face the problem of what to do about it during pregnancy. In my experience, narcolepsy does not affect fertility. The issue is how to treat symptoms during pregnancy. The safest approach is for the pregnant woman to stop using medications during pregnancy, just as she would stop taking most prescription and over-the-counter drugs. If the alternative treatments such as daytime naps and other accommodations are not possible, the pregnant woman may have to take a medical or other leave from employment. Further, it is dangerous and in many places illegal for her to drive while she is off the medication. Pregnant women must also make sure they do not become deficient in iron and folic acid as that may cause them to develop restless legs syndrome, which would worsen their already severe sleep problem.

After the baby is born, the mother or caregiver must be alert when taking care of both the newborn and any other children. Mothers who have stopped taking medications during pregnancy should probably resume them after the baby is born; however, because the long-term effects of these medications on breastfed babies are not known, pregnant women should discuss the problem with their doctor before giving birth. A narcoleptic mother might decide that bottle-feeding is the best option. Additionally, narcoleptic mothers might find that extra help at home is necessary to help them cope. Fathers could take over the bottle-feeding chores at night, for example. New motherhood can be a trying and tiring time for any woman, but it becomes significantly more so when the new mother is narcoleptic.

Back to the First-Year Med Student

The medical student with narcolepsy had legitimate concerns about how the disorder was going to affect her life, and whether she would be able to complete her studies. Although sleeping had been a problem for her for years, she now knew that her sleep disorder was a medical condition that could be treated, though not cured. I encouraged her to go to her doctor for diagnosis and treatment. I also told her about another medical student I had diagnosed with narcolepsy a few years earlier. This student was referred to me when he was in his last year of medical school. He had been falling asleep at rounds and was functioning at a low level. His instructors believed he was lazy; they wanted me to confirm that there was nothing medically wrong with him so they could fail him. Deciding to fail a student who has already spent more than three years in medical school is not a decision taken lightly, and it is considered only when there are severe performance issues. I found that the student was not lazy; he had instead a classic case of narcolepsy, which had been missed by his professors and doctors—who could have caught it by asking him a few simple questions. He was started on treatment, graduated from medical school and finished his postgraduate specialty training, and is now a successful doctor. After hearing this, the medical student was reassured that there was hope for her. The lecture she might have dozed off in had probably saved her career.

Like people with restless legs syndrome and sleep apnea, narcolepsy sufferers can go undiagnosed for years. Even medical professors can miss the implications of its symptoms; the average overworked family practitioner is even less likely to notice them. But while the quality of life may decrease for RLS sufferers, life itself is at risk for people with sleep apnea or narcolepsy. They need to get an accurate diagnosis as soon as possible.

14

Fear of Sleeping and Other Unusual Ailments

THE MYSTERY. Why are some people afraid to go to sleep at night? Can nightmares and uncontrollable, sometimes violent behaviors be cured? We need to learn to distinguish between harmless sleep behaviors and abnormalities that indicate a serious problem.

~~~~~~~~~~~~~~~~~~

The Case of the Woman Who Was Afraid of Her Dreams

In the specialized field of sleep medicine, the best way to discover what is really going on with a patient's disturbed sleep can be an interview with both the patient and the bed partner. The forty-six-year-old woman in my office had come accompanied by her husband, and it was he who provided the most useful information on his wife's sleep patterns. She had been referred to the sleep clinic because she was afraid to fall asleep. When she started to describe her problem, she was almost smiling, as though she were somewhat embarrassed about wasting my time with her silly story. Her husband, on the other hand, seemed extremely concerned, even upset. Her problem was that she had "lots of dreams and bad dreams," which had been plaguing her as far back as she could remember. Even as a young child, her parents would have to come into her room to calm her down after one of these events.

Now, as an adult, she dreaded going to sleep each night because she knew she was likely to have a bad dream. The episodes usually began at about one in the morning, and in most of her dreams, she would be trying to protect herself from a masked man who was trying to stab her. Her husband had been awakened by her dreams, and he was able to observe her reactions. She would yell, turn her head from side to side as though avoiding an attacker, ball up her hands into fists, and strike out. She often hit her husband "pretty hard" while in these dreams, and he had the bruises to show for it. Sometimes her husband would waken her during the more severe episodes, and she would be afraid to go back to sleep. She would try to think pleasant thoughts, and if her husband hugged and soothed her she was sometimes able to drift back into a dreamless sleep. But often she would find herself right back in the same dream. Fear of her nighttime episodes had made it impossible for the couple to travel or stay with other people. She and her husband had experienced this trauma for most of their twenty-eight years together, and now she was desperately seeking relief—for herself and for her husband.

In answer to my questions, she said that she was taking no medications and had never had a brain injury, severe infection, or loss of consciousness. She also had no history or symptoms of psychiatric disorders. For some forty years, she had found sleep a painful experience; it did not refresh her at all. But fortunately her problem was easy to diagnose and a treatment was available that would solve her problem virtually overnight.

The Many Faces of Sleep

Although sleep is normally considered a peaceful activity, it can also be a time when distressing visions and abnormal behaviors occur. Nightmares and terrifying visions can be the result of a number of factors; abnormal behaviors, called parasomnias, occur when the mechanisms that control such behaviors as yelling, screaming, walking, talking, and urinating fail to function or function abnormally while the individual is asleep.

We saw in Chapter 1 that our brains have three states of consciousness: waking, non–rapid eye movement (NREM) sleep, and REM sleep, which is the dream state. For some people, however, the boundaries between the three states can collapse.

When we are awake, both our brain and our muscles are active: we can think and our senses continuously give us information about our environment. Our bodies automatically maintain muscle tone, control our breathing, heart rate, and blood pressure, and make us aware of physical needs such as eating, drinking, and urinating.

During NREM sleep, we continue to receive sensory data from the environment, but the body and brain filter out the irrelevant information. After a few nights of being awakened by a plane flying overhead at 4:00 A.M., we start to sleep through it—but we continue to respond to noises coming from the baby's room down the hall. Our brains continue to control all the automatic functions; sphincters in the body keep various fluids where they belong; and muscle tone is maintained. Some mental activity and some dreaming may also occur during this state.

REM sleep is the state in which we dream, and when we dream our body is almost entirely paralyzed. The muscles of the arms and legs cannot move. The main breathing muscle, the diaphragm, continues to work, as do sphincters in the intestinal tract, but some of the automatic functions which control body temperature, blood pressure, and heart rate might become erratic.

But when a person experiences a breakdown in the boundaries between these states it can result in several disorders, such as sleep terrors, sleepwalking, and reacting physically to violent dreams. Abnormal behaviors during sleep can occur during both NREM sleep (scientists call these NREM parasomnias) or REM sleep (REM parasomnias).

The Dreams and Nightmares of REM Sleep

TERRIFYING AND REPEATED NIGHTMARES
Nightmares are dreams that are frightening and vivid. I find it surprising that people do not complain of these more often, as most people dream three to five

times a night. The reason is that most of the time nightmares are of no consequence, and people either forget them when they wake up or learn to ignore
them. Some people cannot remember any of their dreams; others remember
one or more a night. Women report having nightmares more often than men
(women are probably more willing to report them).

Patients with posttraumatic stress disorder (PTSD) frequently have terrifying dreams that replay the traumatic event they have lived through. They
might awaken from the dreams sweating and terrified, with their heart pounding. Because these experiences are so distressing, some develop a fear of falling asleep. Although PTSD is commonly found among persons who have
served in the armed forces and seen combat, it can occur in any person who
has experienced or witnessed severe physical or psychological trauma. I have
known people who have had these dreams nightly for forty or fifty years.

Persons with PTSD should seek help from medical practitioners. Scientists believe that a contributor to PTSD is activation of receptors in the brain
that also control blood pressure; these nightmares can be treated with an old
drug called prazosin, which was initially used to treat high blood pressure.
Indeed, research in 2016 reported that PTSD sufferers with higher blood pressures were more likely to respond to this treatment.

Children, too, can acquire a fear of going to sleep. When a child has a
nightmare, the parent should demonstrate that the fear is unfounded. If, for
instance, the child thinks there is something hiding under the bed or in the
closet, the parent should look in these places with the child to show that there
is nothing there. In rare cases when the problem persists, parents should seek
help from a doctor or a psychologist.

DREAMING, BUT NOT QUITE ASLEEP

People who are extremely sleep deprived or who have narcolepsy sometimes
dream even before they fall asleep. Sometimes they dream after they have
awakened. This is not normal behavior. People do not usually dream until
they have been asleep for roughly ninety minutes.

As we saw in the previous chapter, the dreams that people have as they are
falling asleep or after they have awakened are called hypnagogic and hypnopompic hallucinations. Sometimes these hallucinations consist of mundane
and fleeting thoughts; sometimes, however, they can be vivid and even frightening. When people are having these hallucinations, they generally know that
the visions are "not real." But they may not realize that the visions are dreams.

The strangest example I ever saw of a person experiencing hypnagogic hallucinations was a young woman who was in an intensive care unit. She had a
severe form of sleep apnea associated with a neurological condition that made
her extremely sleepy during the daytime, and as I approached her, I could
see that she was talking to an invisible object that seemed to be about fifteen
yards away from her. She told me that she was talking to the giant Cheshire cat
"over there." She then turned to me and smiled and said, "Of course I know
there is no Cheshire cat over there. This is just some sort of dream."

The ability to recognize that the hallucination is not real is quite different
from the experience of a patient with schizophrenia who has hallucinations.
The schizophrenic patient believes that the hallucinations are real. The inability to differentiate reality from hallucination is, in fact, one of the hallmarks
of schizophrenia (see Chapter 16).

Dreamlike hallucinations might also occur in people who are awake,
even as they are participating in an activity. I've even had patients who had
hypnagogic hallucinations as they were driving! People experiencing such
hallucinations should stop driving and consult their doctor.

We tend not to treat these hallucinations. Instead we focus on the conditions that cause them. The hallucinations themselves are usually not distressing to adults, though they can be to children. The most likely trigger for this
type of hallucination is sleep deprivation caused by lifestyle, and people who
experience them should consider making lifestyle changes that will enable
them to sleep more. If sleep deprivation is not the cause of the hallucinations,
daytime sleepiness is not present, and the hallucinations are not occurring
during risky activities such as driving, doctors might need only to reassure
patients that the hallucinations are not a sign of something dangerous. I
might suggest REM-suppressing drugs (usually antidepressants) to narcolepsy
patients if the hypnagogic hallucinations are frequent (occurring more than
once a week) and disturbing.

PHYSICALLY REACTING TO DREAMS

Patients who have entered the blurred state between REM and NREM sleep
are not paralyzed, as they normally are during REM sleep, and thus might
find themselves reacting physically to nightmares, thrashing about or kicking.
They have a condition called REM behavior disorder (RBD). People with this
problem, which is more commonly found in men than in women, have been

known to inflict severe injuries on their bed partners, who should encourage the dreamer to seek help for such behavior. Most of the time, the violent activity is related to a dream that includes being attacked by an unknown but terrifying person or animal. One patient of mine dreamt that he was being attacked on the beaches of Normandy during World War II.

Because dreamers with this disorder can react physically, they can be a danger to themselves and bed partners. I had a patient who dreamt he was being chased by a moose and a bear (go figure) and was running away from the animals toward a building. He arrived at the building, but the door was closed, and the animals were coming closer and closer. He started to bang his fists on the building—only to be awakened by his wife's screams. He had been pounding her, not the building. I have had other patients report that they have banged their fists through glass, broken lamps, and damaged furniture. In one case, a patient lunged out of bed, fell on the floor, broke his neck, and died. RBD is a serious condition, and sufferers need to get treatment!

Ninety percent of the people who have RBD are men (though we do not know why), but women need to be aware of the disorder as well since they are likely to be secondhand sufferers from it: two-thirds of men who suffer from RBD have assaulted and often injured their spouses. People who have had head trauma or an infection of the brain earlier in their lives are more likely to get the disorder, and it is more common among alcoholics. It has also been reported as a rare complication of certain antidepressants. Some people with this condition may go on to develop Parkinson's disease or other severe neuro-logical conditions years or decades later.

Medical science has not known about RBD for very long: it was first described in 1987. People with RBD are often embarrassed to talk about it, or they are inhibited by fears that they might have a psychiatric condition. Although RBD bears some similarities to PTSD, the two are very different. Most PTSD patients have terrible recurrent nightmares, but they cannot react physically to what they are dreaming because the dreams occur in REM sleep and they are paralyzed. However, some PTSD patients will also have RBD, and they will react physically to their dreams. People who react physically to violent dreams, especially if they have injured themselves or others, should seek medical help. Some patients (and their bed partners) have suffered for forty or fifty years, yet RDB can often be treated effectively with clonazepam, an anti-epilepsy medication, and melatonin, a hormone (see Chapter 20).

SLEEP PARALYSIS

In this disorder, which usually affects adults, the sleeper wakes up from a dream and finds that he or she cannot move. Sleep paralysis can be quite frightening, especially the first few times it happens. Sometimes the sensation of paralysis occurs while the sleeper is actively dreaming, and if the content of the dream is frightening, the sensation can increase the fear. A sleeper might dream that there was someone in the room or that something unpleasant was happening in the house, such as a robbery. I have treated women who dreamt of a devil-like creature that was about to sexually violate them and men who dreamt that they were about to be raped by a creature resembling an old woman. Sexually disturbing dreams like these have been described in several countries around the world. Episodes of sleep paralysis might last only a few seconds or perhaps a few minutes. What sometimes snaps the person out of the paralysis is being touched by another, but the sleeper can do nothing to stop the paralysis. It goes away on its own.

Sleep paralysis is another disorder caused by the blurring of the boundary between wakefulness and REM sleep. The sleeper's brain is awake, but one of the manifestations of REM, the paralysis, remains.

Sleep paralysis is a feature of narcolepsy, but it can also occur in people who are experiencing severe sleep deprivation. I have seen some cases of sleep paralysis that ran in families. Although sleep paralysis in itself is not danger-ous, if the patient finds it distressing, I usually treat it in the same way I treat sleepwalking. If reassurance does not help, in more severe cases I might pre-scribe a drug such as clonazepam or one of the antidepressants that suppress REM sleep. People experiencing disturbing sleep paralysis should consult with their doctor.

Abnormal Behaviors in NREM Sleep

SLEEPWALKING

Sleepwalking occurs when parts of the brain are asleep and other parts, the ones that control walking and other physical activities, are in some way awake. The part of the brain responsible for thinking and alertness is asleep, and sleepwalkers usually have no recollection of sleepwalking after they wake up. Although for humans it is unusual to have the brain be awake and asleep at the same time, it is a common state for some other animals. Certain marine mammals—for example, dolphins—can continue to swim around while one side of their brain is asleep and resting because the other side of their brain is wide awake and controlling various functions. This ability may be what allows marine mammals to spend their whole lives in the water.

As many as 10 to 15 percent of Americans have sleepwalked at some time, particularly when they were children. Sleepwalking becomes much less common as people leave the teenage years, although I have had several adult sleepwalking patients. While sleepwalking, the person gets out of bed and starts walking, demonstrating what is best described as robotlike behavior. The walking might seem purposeful—for example, the sleepwalker might go to the kitchen—but it generally is not. One of my sleepwalking patients, a child, would walk into the laundry room and urinate into the laundry hamper.

Sleepwalking seems to occur most often during very deep (slow wave) sleep. Children spend more of the night in deep sleep than adults do, so it follows that they sleepwalk more often than adults. Deep sleep is also more common in the first third of the night, which is when sleepwalking is most likely to occur. Adults and children who are sleep deprived also fall into a deep sleep more quickly and tend to sleepwalk more often. Sleepwalking seems to run in some families and is also found more often in people who are under stress or who have been drinking alcohol. Defendants accused of violent crimes have offered the defense that they were sleepwalking and sometimes even been acquitted, though it is difficult to prove what state a person was in when the crime was committed.

Most of the time, sleepwalking is not dangerous unless the walker ventures outside or turns on appliances (such as a stove). Usually, he or she returns to bed, still asleep. In cases where the sleepwalking is not associated with dangerous actions, nothing needs to be done. People encountering a family member sleepwalking should not waken the person, but lead him or her back to bed. Sleepwalkers who are awakened too abruptly might be upset and have trouble falling asleep; they also might become overly concerned about what their sleepwalking signifies. If the sleepwalker has been discovered in a dangerous situation, then the chance of harm must be reduced. Alarms could be installed for a sleepwalker who has stumbled down the stairs, for instance. It's a mystery to me (though a good thing) that people do not hurt themselves more often when they are in this state.

Patients whose sleepwalking results from sleep deprivation, stress, or alcohol abuse can address this with their doctor. If sleep deprivation is the culprit, getting proper amounts of sleep usually solves the sleepwalking problem — in severe cases of sleepwalking, the sleepwalker might not know the best techniques for getting the right amount of sleep, and a sleep doctor can help. If the sleepwalking occurs on nights when the person has been drinking, the problem might be solved by eliminating alcohol consumption. When the sleepwalking seems to be related to stress, the doctor should try to help the sleepwalker find and eliminate the cause of the stress. This might involve referring the patient to a psychologist.

Because these treatments are usually effective, I seldom recommend medications for sleepwalking unless they can suppress sleepwalking for patients who have had dangerous episodes. I had one patient, for example, who found herself walking in a cemetery several blocks from her house, wearing only her nightgown. She was barefoot, the temperature was below freezing, and there was snow on the ground. I will also sometimes recommend medications when the patient is traveling; sleepwalking in a strange environment can be dangerous. The medication I recommend most often is clonazepam, but sleepwalkers should discuss all medications with their doctor to determine whether a particular one is right for them.

SLEEP TALKING

Sleep talking is quite common among adults and children. Most of the "talk" is gibberish, although a listener might be able to make out individual words. I have not heard of people blurting out secrets during sleep talking episodes. This might be considered an embarrassing condition, but it is not one that requires treatment.

SLEEP TERRORS

This disorder, which can occur in both children and adults, is also called "night terrors." Sufferers get out of bed abruptly, sometimes screaming with their eyes wide open, sometimes sweating. They appear to be terrified, and some look as though they are about to commit a violent act.

Although the sleeper might appear to be reacting to a dream, usually he or she is not. Sleep terrors are a form of sleepwalking, and the treatment is the same. There is no need to awaken people who are having these episodes; it is best to calmly put them back to bed. The following morning, they usually have no recollection of the event. Sleep terrors are a bizarre form of behavior, but they are rarely dangerous enough to require treatment.

URINATING IN BED

Enuresis, or urinating in bed, occurs when the mechanisms that keep the sphincters of the urinary system working fail to function properly. This is a problem mainly found in children (and twice as often in boys than in girls) and the elderly. In children, the problem is caused by slow development of bladder control. In the elderly, it is generally related to changes in anatomy brought on by the aging process or is a symptom of a disease.

Childhood enuresis can be very troubling for both the child and the parents. Children might develop a fear of going to sleep or sleeping at a friend's house because they are afraid that they will wet the bed. Parents should take a child with this problem to be evaluated by a pediatrician in case there is a medical reason (for example, sleep apnea or a urinary tract infection) for it. The pediatrician can then advise the family about what to expect and how to handle the problem. In many cities, there are specialized clinics that deal with this situation.

If there is no medical problem, parents can try using an alarm system that is triggered when the bed gets wet. The alarm wakes up the child, who eventually learns bladder control. If this treatment is not effective, the doctor might recommend one of several medications. Desmopressin acetate (DDAVP) is a medication that imitates the effect of a chemical produced by the pituitary gland that reduces the amount of urine. This drug, which is immediately effective, can be taken just before bedtime either in a nose spray or in pill form. A low dose of the antidepressant imipramine taken one to two hours before bedtime has been used for many years to treat children who wet the bed, but this medication is successful less than half the time. These treatments do not cure the problem, though. Usually the problem is solved as the child develops and gains greater bladder control.

Incontinence can also become a problem for people as they age. Urinary tract infections, diabetes, diseases of the prostate in men, and vaginal infections in women may play a role. About one in twenty older women (over sixty-five) worldwide wets the bed at night. If it is not possible to solve the medical problem causing the incontinence, then the only solution may be to use an incontinence pad. If there is no medical problem, then exercises that help tighten the muscles used in the control of urination (called Kegel exercises) might be helpful.

NIGHT SWEATS

Excessive sweating during sleep can be extremely distressing and embarrassing. I have had patients who developed a fear of sleeping because of the discomfort caused by waking up with drenched sheets and pillows. Patients who experience severe night sweats should consult a doctor: such sweating can be associated with menopause, but it can also be a problem for patients with sleep apnea, restless legs syndrome, hyperthyroidism, certain infections, and cancers. In some cases doctors are unable to find the cause.

TEETH GRINDING (BRUXISM)

Bruxism is an increase in the activity of the jaw muscles during sleep. This condition can be found in children and adults; it is equally common in females and males, and it is more common among people who are heavy users of tobacco and alcohol. Scientists do not know much about it, but it is more commonly found in people under stress. It can also occur as a reaction to certain drugs. The grinding can wear down the sleeper's teeth, as well as being disruptive for bed partners. For some sufferers, stress reduction will also take care of the bruxism. But if the sleeper's teeth are wearing down, or if he or she experiences pain in the jaw, the tooth grinder should consult a dentist, who might recommend a mouth guard to be worn at night.

HEAD BANGING AND BODY ROLLING

One of the most unusual problems we see in the sleep clinic is a disorder in which a sleeper repeatedly bangs his or her head against a mattress, a crib, or a wall. Some people rock their bodies throughout the night. Worldwide, the disorder is found in about 10 percent of seven-year-olds, though the number decreases among older children, and most children grow out of it. This condition is four times more common in males than in females, although we do not know why. It can be frightening to see, especially for parents, but this is not a serious problem. The patients are fine when they wake up. Since some people with neurological problems display similar movements, it is a good idea to consult a doctor. If the cause is not neurological, we do not normally treat the disorder unless the person is in danger of self-injury.

Back to the Woman Who Was Afraid of Her Dreams

The woman who had dreamed almost nightly for forty years that she was going to be stabbed by a masked man had REM behavior disorder. I could not determine the cause of her problem because she had no history or evidence of brain damage, had never been in a coma or lost consciousness as far as she knew, was not an alcoholic, and was not on any medication that might lead to this problem. I recommended to her family doctor that she be started on clonazepam, which is also used to treat some forms of epilepsy and panic disorder. It is sometimes prescribed to help patients fall asleep because it seems to make the brain less likely to respond to stimuli, including the stimuli from dreams. When I spoke to my RBD patient several months after she began taking the medication, she was dramatically improved. She no longer feared dreaming about the killer. Her husband told me that she sometimes still moved a great deal during sleep, but the hitting had stopped.

The disorders mentioned in this chapter are not usually dangerous to the sleeper, but sufferers from them might want to see their doctor, to ensure that these are not symptoms of a more serious disorder. The exception is REM behavior disorder, which generally is more dangerous to the bed partner, although it can be debilitating for the patient, who is not getting enough of the right kind of sleep. The disorder is more common among males, but women are frequently the victims of the physical assaults and violence that accompany it. People exhibiting violent behavior in bed need to see a doctor or sleep specialist about treatment.

15

Medical Conditions That Affect Sleep

THE MYSTERY. Why do so many medical disorders, including hormonal abnormalities, diabetes, cardiovascular diseases, and arthritis, result in sleep disorders?

~~~~~~~~~~~~~~~~~~~~~~~~

The Case of the Woman with a Black Curtain

The woman sitting in front of me was in her seventies, articulate, and obviously distressed. She had just walked perhaps ten yards from the waiting room to my office and she was breathless, almost panting. I let her settle down, and when she had composed herself, I started to interview her, expecting her to focus on sleep issues—after all, she was at a sleep clinic! Instead, she told me that a black curtain had descended over her mind, and because of it she could no longer continue her work as a visual artist. She was trying to prepare for an art show, but she was unable to work. Although she admitted to having trouble sleeping, she insisted that her main problem was that her creativity had been blocked by the black curtain.

Her family doctor, on the other hand, believed that her main problem was the insomnia and had referred her to the sleep clinic. I examined the woman and found that she had swollen ankles and noises in her lungs. This told me that an entirely different organ system was causing her insomnia, and I knew I would be able to help her lift the black curtain.

Chronic Medical Problems Can Lead to Sleep Problems

Insomnia is often a symptom of a disease. When a person is suffering from insomnia, he or she should consult a doctor, who needs to determine the cause of the symptom and then treat it. In addition to the triggers discussed in Chapter 10, insomnia can be caused by many different medical problems. A disturbance in sleep often indicates that there is something else wrong in the body, and the problem might be serious. Disorders involving almost every single organ system can cause problems with sleep. This is why it is imperative for doctors to ask their patients about how they are sleeping (though until recently they seldom did) and for patients to describe sleep problems when they see a doctor.

Name any chronic disease, and you will find that it is probably associated with a sleep problem. Diabetes, kidney failure, arthritis, Parkinson's, heart failure, and cancer are some of the commonly occurring medical conditions that can affect sleep and lead to sleep complaints. And a host of psychiatric complaints— depression, bipolar disorder, obsessive-compulsive disorder—also have insomnia as an important symptom. (These will be reviewed in Chapter 16.) Diseases and disorders of the nervous, pulmonary, cardiovascular, and urinary systems and the gastrointestinal tract, imbalances and problems affecting the sinuses and hormones, and some other conditions can all cause sleep problems. A variety of sleep problems are also associated with cancer.

Diseases and Disorders of the Nervous System

The nervous system is made up of the brain, the spinal cord, and the nerves that go to all parts of the body, including to the organs, muscles, and skin.

Because the system that controls sleep is in the brain, disorders of the nervous system are likely to result in disturbed sleep.

ALZHEIMER'S DISEASE

Alzheimer's is a neurodegenerative disorder in which the nervous system deteriorates, usually owing to an accumulation of beta-amyloid, a chemical in the brain that forms plaques, and strands of a protein called tau, which appear to choke off normal brain cells and their ability to communicate with each other. Inflammation, perhaps as a response to infection, may play a role. The disorder can progress very rapidly or slowly, with a gradual loss of brain function (including the ability to sleep) over a period of many years. Lewy Body Dementia, a less common disease, shares some of the symptoms of Alzheimer's. Cognitive decline and memory problems are the main symptoms.

The statistics concerning Alzheimer's are staggering. In the United States, it was estimated in 2016 that there were 5 million people living with the disease and that by 2050 that number will be about 14 million people. It has been estimated that one in nine of all Americans over age sixty-five, and perhaps half of people older than eighty-five, have Alzheimer's. Because women on average live longer than men, many more women have Alzheimer's than men. It has also been suggested that within a given age group, a woman is more likely to have Alzheimer's than a man. A study from Denmark estimated that of a hundred women age ninety or older, eighty had Alzheimer's; in contrast, of a hundred men age ninety or older, twenty-four had Alzheimer's. Besides the primary symptoms of the disease, 42 percent of Alzheimer's patients are depressed.

Recent research suggests that sleep apnea may contribute to the cognitive decline of Alzheimer's patients. Thus, treating their sleep apnea (see Chapter 12) might slow cognitive decline. This is an exciting development. People with severe Alzheimer's may spend a great deal of time awake at night, and many may reverse their days and nights, sleeping fitfully throughout the day and remaining awake all night. In addition, some Alzheimer's patients may experience sundowning, a condition that usually occurs in the late afternoon or early evening, in which the patient becomes agitated and has hallucinations or episodes of anxiety. It is believed that this symptom might be related to a breakdown of the circadian clock or be due to poor quality sleep. Sundowning is common among elderly people in general, not solely Alzheimer's patients. About one in five institutionalized older people experiences sundowning.

New medications are available that seem to slow the pace of Alzheimer's and might result in some improvement in brain function. These drugs can cause nausea, vomiting, diarrhea, and insomnia, however, and their effect on Alzheimer's patients' sleep patterns has not been widely studied. Sometimes patients are given other medications, such as antidepressants, that also might cause insomnia. Recent research suggests that exposure to light during the daytime could be helpful in improving nighttime sleep because it helps reset the patient's circadian clock. Some scientists have recommended the use of melatonin, a hormone that is sometimes recommended for sleep problems (see Chapter 20). Several studies have suggested that melatonin at a dosage of 6 to 9 milligrams may be effective in improving sleep and treating sundowning in Alzheimer's patients.

HEADACHES

Headaches can be associated with sleep and other medical disorders. Some people with sleep-breathing problems awaken with headaches; it is common, for example, among people with sleep apnea. Not breathing enough causes an increase in the carbon dioxide level in the blood, and this leads to an increase in blood flow to the brain, putting extra pressure in the brain. Some patients have severe headaches that always begin during sleep. These can occur infrequently or almost every night. Little is known about these headaches, though some reports suggest that such headaches can be treated with caffeine, or lithium (a drug used to treat bipolar disease), flunarizine (a drug used to treat migraines), indomethacin (an anti-inflammatory), or Topamax (an anti-epilepsy medication). Some patients do not respond to any treatment.

Headaches are much more common in women than men. The only exception is the cluster headache, a less common type of headache, which is diagnosed more often in men. People with severe headaches often have difficulty falling and staying asleep. Two types of severe headaches that typically have an effect on sleep are migraines and cluster headaches.

Migraines. Migraine headaches, which are about three times more common among women than men, can be incapacitating. These headaches often affect only one side of the head and consist of a throbbing pain, sometimes accompanied by nausea and vomiting and/or increased sensitivity to light, sounds, and smells. For many people, migraine attacks are recurrent, though they tend to become less severe with aging. Many people begin to see shimmering lights around objects, zigzag lines, and wavy images about ten to thirty minutes before the onset of a migraine. Some experience hallucinations or even lose their vision temporarily. Not only do these headaches interfere with sleep, but also the ensuing sleep loss may cause the migraines to become more frequent. I recommend that people with severe migraine headaches (indeed, any severe headache) be evaluated and treated by a neurologist specializing in headaches.

Cluster headaches. The cluster headache is perhaps the most severe form of headache. The headache episodes occur in clusters or one after another within a short period of time, often lasting two to four months. During one of these clusters, the person might have two to ten headaches a day, often occurring at the same time each day like clockwork until the cluster ends. A period of several months or even years might follow before the next cluster begins.

The cluster headache is on one side of the head and face and often begins with a drooping of the eyelid, tearing, and enlargement of the pupil of the eye on the affected side. The pain becomes unbearable after five to ten minutes. The headache usually lasts about thirty to forty-five minutes (although some last up to two hours). Once the headache starts to subside, the pain might dissipate within five to ten minutes. The headaches are so severe that people pace, rock their bodies, and sometimes bang their heads against a wall to try to stop the pain. These headaches commonly start during sleep, particularly dreaming (REM) sleep, and frequently the pain awakens the sufferer. People with cluster headaches should seek medical help, though because some doctors consider this type of headache a male phenomenon—it is roughly twice as common in men—they often miss it in women. Women experiencing this disorder should describe their symptoms as precisely as possible. These headaches sometimes respond to breathing oxygen, as well as to some of the medications used to treat migraine headaches.

PARKINSON'S DISEASE

Parkinson's disease is a common neurological problem that affects the parts of the brain that produce the chemical dopamine. Parkinson's affects about 1 million Americans and 10 million people worldwide. The disease causes involuntary movements including tremors, masklike facial expressions, and an abnormal gait. Parkinson's disease occurs more frequently in older people, and is 50 percent more common in men than women.

About 60 percent of people with Parkinson's disease have trouble falling and staying asleep. Experiencing restless legs syndrome (see Chapter 11) and repetitive twitches in their legs as they sleep is common among these patients. They might awaken during the night and be unable to fall asleep again. Sometimes this is the result of their medication wearing off and being eliminated from the body. Sometimes days and nights become reversed.

About a third of patients have REM behavior disorder, a condition in which the sleeper physically reacts to the content of his or her dreams (see Chapter 14). Patients with this condition can harm themselves or their bed partners. Many patients also experience frightening hallucinations while they are awake. As a result of RBD, sleep apnea, a movement disorder, or the medications used to combat the disease, patients can experience severe daytime sleepiness. Some doctors have begun to treat this severe daytime sleepiness with the wakefulness-promoting medication modafinil. When taken in the morning, modafinil does not interfere with a patient's ability to sleep at night (see Chapter 20).

Diseases and Disorders of the Pulmonary System

Any lung problem, most commonly asthma and chronic obstructive pulmonary disease (COPD) that causes excessive coughing, wheezing, or shortness of breath at night, can lead to insomnia.

ASTHMA

Asthma, which is the constriction of the bronchial tubes causing excessive coughing and shortness of breath, affects roughly 8 percent of the U.S. population. Until they reach thirty years of age, men are more likely to have asthma than women. After age thirty, it is about twice as common among women than men. This is important because when the disease starts at a younger age it tends to improve, while in the older age groups in which women predominate, it is much less likely to improve. Female asthmatics have a 70 percent higher risk of being admitted to the hospital for asthma than their male counterparts.

Medication usually relieves the symptoms of asthma. For some people, however, the first sign that their asthma is not under control is when they develop a wheeze that awakens them from sleep. Sometimes a sleeper who is awakened by coughing does not understand what is happening. Some patients who are referred to me because of insomnia discover the extent to which the coughing disturbed their sleep only after they have had a sleep test.

CHRONIC OBSTRUCTIVE PULMONARY DISEASE

Chronic obstructive pulmonary disease (emphysema) occurs in long-time cigarette smokers. It is most often irreversible and can be fatal. Research suggests that tobacco has a greater adverse effect on lung function in women than men, and women are more likely to be hospitalized for COPD. As with asthma, coughing and shortness of breath are common symptoms of COPD. In addition, because many COPD sufferers continue to be cigarette smokers, they often wake up during the night craving nicotine. Some people awaken in the morning with their lungs filled with sputum, which they have to cough up in order to breathe normally. The combination of waking up with coughing and shortness of breath makes it difficult for these patients to have restful sleep.

Diseases and Disorders of the Cardiovascular System

Although there is a widespread belief that diseases of the heart and blood vessels affect mainly males, this is not true. In fact, about sixty thousand more women than men die each year of cardiovascular disease in the United States. In 2015, about 80 million Americans were estimated to have cardiovascular diseases. Recent research has shown that cardiovascular diseases can cause sleep problems—and sleep disorders can cause cardiovascular diseases.

ANGINA

Angina occurs when the muscles of the heart suddenly, but temporarily, stop receiving enough blood flow, but the muscle is not damaged. The main symptom is chest pain brought on by exertion, which is relieved within minutes of stopping the exertion (such as exercise) or a medication that dilates the coronary arteries. Patients with coronary artery disease may develop symptoms of angina when they sleep, if they have sleep apnea, or sometimes during dreaming sleep, when the arteries to the heart may go into spasm. Patients with this symptom should report it to their doctor.

HEART ATTACK

A heart attack (what doctors call a myocardial infarction) occurs when the muscles of the heart are suddenly deprived of blood flow, and the muscle is damaged. Diseases of the blood vessels of the heart (the coronary arteries) are the most common cause. Women who are having a heart attack can have different symptoms from men. Although chest pain is considered the classic symptom of a heart attack, research reported in the United States in late 2003 found that 43 percent of women did not have chest pain during a heart attack. However, 70 percent of women who had a heart attack suffered from fatigue and about half had had trouble sleeping in the weeks before the episode. Fatigue was the most common symptom.

HEART FAILURE

Heart failure (often called congestive heart failure, CHF) occurs when a weakened heart muscle cannot pump enough blood to meet the body's requirements for blood and oxygen. There are many causes; the most common is having had a heart attack that has permanently damaged the heart. About 6 million Americans suffer from some form of heart failure. Although this has commonly been considered a problem affecting men, statistics indicate that heart failure is a bigger problem for women. In 2016, about 3 million women in the United States were living with heart failure compared to 2.7 million men. More women than men die of heart failure.

Some people with heart failure develop a sleep-breathing pattern that at times becomes progressively deeper, then progressively shallower. Sometimes breathing stops completely for short periods of time. To restart the breathing the brain has to go through a mini-awakening. When the pattern of breathing too much followed by breathing too little repeats itself about once a minute it is a sign of sleep apnea, which results in insomnia. People with heart failure frequently have trouble falling asleep, and when they awaken during the night they are frequently extremely short of breath and feel as though they must sit up. Some of the medications used to treat heart failure, such as water pills and diuretics, may result in frequent trips to the bathroom at night, which also disrupt sleep.

People with heart failure might face the unfortunate combination of extreme sleepiness (because their sleep is so disrupted) and an inability to fall asleep. Research is still ongoing to evaluate oxygen and breathing devices that normalize the breathing pattern in these patients.

PALPITATIONS

Some people awaken during the night with an abnormal heart rhythm. They might notice that the rhythm seems very fast or very slow, or they might feel an irregularity, such as an extra or missed beat. They might notice that they have to urinate at night. Some people are woken up by frightening dreams; in these cases, the palpitations might not be a medical problem. In other cases, however, the symptom could represent a significant cardiac arrhythmia that should be evaluated by a doctor.

Some people awaken from sleep with a rapid heartbeat, sweating, and the feeling that they are about to die. Sometimes this feeling of impending doom is caused by a panic disorder, which for some occurs mostly at night; it can also be a symptom of posttraumatic stress disorder (see Chapter 16). The episodes can be so alarming that the sufferer could develop a fear of falling asleep.

HIGH BLOOD PRESSURE

The heart pumps blood in arteries to the rest of the body. The amount and speed of blood the heart pumps and the resistance to blood flow in the arteries determine the pressure inside the arteries. This resistance could be increased by diseases such as atherosclerosis. In order for blood to pass through restricted arteries the pressure in the arteries is increased. Long-term increase in blood pressure is called hypertension, a major cause of heart disease, stroke, and kidney disease, which are all associated with sleep issues. When blood pressure is measured, two numbers are recorded: systolic blood pressure (the normal is up to 120) is the pressure while the heart is pumping; diastolic pressure (the normal is up to 80) is the pressure while the heart is relaxing. Thus a normal blood pressure would be 120/80. Hypertension is defined when either the systolic pressure is greater than 140 or the diastolic pressure is greater than 90. Hypertension is very common, and increases with age. In 2016 it was estimated that about a third of Americans aged forty-five to fifty-five and about two-thirds aged sixty-five to seventy-four have hypertension. Hypertension is generally treated with medications and lifestyle modifications such as diet and exercise.

Scientists have now shown that sleep apnea can cause high blood pres-

sure; it can also make control of blood pressure with medications more difficult. Some medications used to treat high blood pressure, such as beta blockers, may cause insomnia and nightmares.

Diseases and Disorders of the Urinary System

The urinary system is made up of two parts. The kidneys maintain fluid bal-
ance in the body, keep electrolytes (such as sodium and potassium) at safe
levels, and remove toxins while producing urine. The second part consists of
tubes that carry the urine from the kidneys to the bladder, which stores the
urine. When the bladder is full and the person urinates, the fluid goes from
the bladder via another tube called the urethra before leaving the body. In
males, in front of the urethra near the bladder is the prostate gland. Quite a
journey!

When a person reduces his or her fluid intake, the kidneys try to keep
water in the body, and urine becomes concentrated, but for sufferers from
certain kidney diseases (diabetes, failing kidneys), the urine does not become
concentrated. The urinary systems of these people consequently produce too
much urine and they have to go to the bathroom many times a night, disrupting sleep. If kidney function fails and the patient has to go on dialysis,
he or she frequently will experience very severe movements or restless legs
syndrome, which inhibits sleep. Menopausal women and older men (especially if their prostates are enlarged) need to go to the bathroom more often at
night and may have a great deal of difficulty falling asleep again. People with
untreated sleep apnea frequently have to urinate at night.

Diseases and Disorders of the Gastrointestinal Tract

There are several common diseases of the gastrointestinal tract that disturb
sleep.

GASTROESOPHAGEAL REFLUX (GER)

At the bottom of the esophagus (the tube that carries food from the mouth into
the stomach) is a sphincter that keeps stomach acid from entering the esophagus. Sometimes this sphincter does not work properly, and acid backs up into
the esophagus (a process known as reflux) and causes heartburn. In the most

severe situation, called gastroesophageal reflux disease (GERD), the acid can damage the esophagus. Gastroesophageal reflux is a common condition that affects men and women equally; scientists estimate that perhaps 30 to 40 percent of the adult population in the United States suffers from a reflux disorder. Women who are overweight or pregnant women have a higher risk for reflux.

When GER occurs at night, it can keep the sufferer from falling or staying asleep. Gastroesophageal reflux can affect sleepers in a variety of ways. The acid might make its way all the way up to the mouth and awaken the sleeper with a bitter taste or with severe coughing and choking. If the acid touches the vocal cords, they can sometimes go into spasm, and the person will feel unable to breathe and as though he or she were going to die. Additionally, GER can cause heartburn, which often awakens the sleeper. Research has shown that even when the acid enters the esophagus without causing pain, it might still awaken the sleeper.

Twenty years ago, there were few effective medications for gastroesophageal reflux. Today doctors can prescribe excellent medications that shut down the stomach's production of acid. These drugs fall into two main categories: blockers of a type of histamine receptor—for example, famotidine, ranitidine, cimetidine, which are now available over the counter—and a class of drugs called proton-pump inhibitors that includes rabeprazole (AcipHex), esomeprazole (Nexium), lansoprazole (Prevacid), and omeprazole (Prilosec).

PEPTIC ULCER DISEASE

When the stomach produces too much acid or cannot properly deal with the normal acids it produces, an ulcer can form in the stomach or in a tube called the duodenum. Often the disease is caused by an infection from a bacterium called *Helicobacter pylori*. Some medications, such as certain nonsteroidal anti-inflammatory drugs, even when used as prescribed, can cause ulcers. People with peptic ulcer disease frequently awaken one or two hours after going to sleep with either pain or the sensation of hunger. Eating food or taking antacids often relieves the pain temporarily. Peptic ulcer disease can cause serious complications, such as bleeding in the intestinal tract. People experiencing severe pain that wakes them from sleep should seek medical help. Excellent treatments, such as antibiotics and proton-pump inhibitors, are available for peptic ulcer disease, and these can also help the sufferer regain a normal sleep pattern.

Sinus Disease

Over the years, I have seen many people who start coughing after they lie down. Frequently they might have a cold or a sinus infection before the cough begins, but they do not cough during the day. As soon as they lie down, however, they get coughing fits. It is likely that the lying down causes their sinuses to drain and some of the secretions may make contact with the vocal cords, leading to the coughing. For some people, this problem might go on for months, and sufferers might be treated for a lung disease such as asthma before the true culprit—draining sinuses—is discovered. In the meantime, their coughing causes them to lie awake or wake up during the night, which affects the quantity and the quality of their sleep.

Diseases and Disorders of Hormone Production

DIABETES

Most diseases involving hormone production can cause disturbed sleep. Some of these diseases are common, and some are much more common among women than men. One of the most frequently found is diabetes.

When a person has diabetes, either the body doesn't produce enough of the hormone insulin or there is a resistance to insulin's effect on cells of the body, resulting in high blood sugar levels. In 2012, 9.3 percent of the U.S. population, 29.1 million people, had diabetes. The prevalence of diabetes varied according to ethnic and racial group: diabetes was present in 7.6 percent of non-Hispanic whites, 9 percent of Asian Americans, 12.8 percent of Hispanics, 13.2 percent of African Americans, and 15.9 percent of American Indians/Alaskan Natives. Research has shown that about 50 percent of people with diabetes experience sleep problems. There are many reasons why they develop these difficulties. First, when blood sugar is too high, the kidney filters the sugar into the urine, which forms more urine than normal and results in an increased need to go to the bathroom, disrupting sleep.

In addition, diabetics might develop blood sugar levels during the night that are too low, which will cause them to awaken with sweating, hunger, and a rapid heart rate. This symptom usually occurs when they have taken too much insulin or have eaten too little food before bedtime.

Patients with the most severe forms of diabetes develop nerve damage called neuropathy. This may cause excessive movements or unpleasant sensations in the legs, such as restless legs syndrome, or pain, all of which keep the sufferers from falling asleep. In addition, the neuropathy might affect the nerves of the gastrointestinal tract, and some patients with diabetes suffer from diarrhea at night, which also affects their sleep adversely.

DISEASES AND DISORDERS OF THE THYROID GLAND

The thyroid gland is located in the neck in front of the trachea (the breathing passage) right below the Adam's apple. This crucial and sensitive gland produces thyroid hormone, which is involved in the regulation of metabolism in most of the body's cells. All thyroid disorders are five times more common among women than men.

Goiter. In some people, the thyroid gland enlarges to such an extent that it starts to block the breathing passage behind it. Most of the time, this is due to enlargement of the cells, sometimes caused by a deficiency of iodine. This condition is called a goiter. When the gland becomes too large, and the breathing passage is significantly blocked, the person might develop sleep apnea. Surgery is often used to remove such an enlarged thyroid.

Hypothyroidism. Sometimes the thyroid gland does not produce enough hormone. This condition, called hypothyroidism, may come on over a period of months, even years. The skin becomes coarse and dry and the hair might start to fall out. The sufferer will gain weight, sometimes becoming clinically obese (BMI of more than 30). The weight gain is caused by a drop in the metabolic rate due to lack of thyroid hormone. With this condition, the person will eat the same amount of food as usual, but because fewer calories are burned, his or her weight increases. Hypothyroidism is at least twice as common among women as men.

Important symptoms of thyroid deficiency include feelings of fatigue (muscle weakness) and sleepiness, which can be debilitating. When the disease is severe, patients might actually lose consciousness and develop a breathing problem that requires immediate treatment. One of the reasons patients who are deficient in thyroid hormone suffer from sleepiness might be sleep apnea. This form of apnea usually develops because the thyroid patient's tongue becomes so enlarged that it blocks the breathing passage during sleep.

The ideal treatment for these patients is thyroid replacement medication: a hormone that replaces the hormone their bodies fail to produce. This hormone replacement medication is usually very effective. Most people with hypothyroidism are prescribed a low dose of hormone to start because their metabolic rate is so low that a normal dose of thyroid might bring on symptoms of hyperthyroidism (see below). Often the dosage is slowly increased over a matter of weeks or months. It is vital that the patient see a doctor regularly to monitor the dosage. I have seen many patients at the sleep clinic who have low or deficient thyroid levels because they have not had their doses increased. Without the correct dosage they continued to have symptoms of low thyroid levels.

Hyperthyroidism. Hyperthyroidism is another major malfunction of the thyroid gland. Like the other thyroid conditions, it is far more common (five to ten times) among women than men. In this disorder, too much hormone is produced and the person becomes hypermetabolic—burning up more calories than he or she takes in. People with hyperthyroidism sweat, tremble, and lose a great deal of weight in a short period of time. Sometimes the disease will cause their eyeballs to bulge outward, a condition known as Graves' disease. Patients with excessive thyroid hormone may find it very hard to fall or stay asleep. They frequently sweat at night and have nightmares that wake them. These patients are sleepy during the daytime and they become physically exhausted because the excess thyroid hormone can reduce muscle strength. Furthermore, hyperthyroidism can trigger an extremely rapid heart rate, which can cause symptoms such as dizziness and fainting. The rapid heart rate can also awaken the person during sleep. Hyperthyroidism is a serious medical condition that should be treated and monitored by a medical professional.

DISEASES OF THE PITUITARY GLAND

The pituitary gland is a pea-sized gland in the brain. This gland produces certain hormones and regulates others that are produced in glands in other parts of the body. There are two abnormalities of the pituitary gland that can affect sleep, acromegaly and tumors.

Acromegaly. In acromegaly the pituitary gland produces too much growth hormone. The effect of the overproduction varies depending on whether the patient has already stopped growing. A child with this condition might become extremely tall; the condition is called gigantism. People with this condition have a characteristic look. Besides being extremely tall, they have larger than normal jaws and foreheads owing to the excess growth hormone. When a person who has stopped growing begins to produce excess growth hormone, some parts of the body might begin to grow again—for example, the jaw and other parts of the face and even the hands and feet. If the disease is not controlled, it can have devastating effects on the body. The heart can become too large and fail. Severe arthritis might follow. Sleep problems can develop. Acromegaly patients have enlarged tongues, which can obstruct their breathing passage when they sleep, leading to severe sleep apnea.

I once had a patient who had been successfully treated for acromegaly with surgical removal of the part of her pituitary gland producing the growth hormone. She had been excessively sleepy for years before her doctor realized that the acromegaly had caused sleep apnea, which now needed treatment. This disorder can also lead to diabetes and might also cause other hormonal problems, especially if the cause of the excess hormone production is a tumor.

Tumors. When tumors of the pituitary become too large, they can squeeze the normal parts of the gland and compress another important area of the brain called the hypothalamus. When this happens, the system responsible for regulating sleep and wakefulness might not work properly, and the person could experience severe sleepiness or develop a random sleep pattern, falling asleep at inappropriate times. But while the tumor is still growing, it is often difficult to diagnose the problem. If it becomes large enough to compress some of the nerves involved in vision it might impair the patient's peripheral vision, causing him or her to bump into things. Doctors can be alert to this symptom. Pituitary tumors can also compress the normal tissue in this gland, which can in turn reduce the secretion of other hormones, including the sex hormones. Another sign that doctors look for in diagnosing and treating pituitary diseases is a reduction in the amount of hair in the pubic area and the armpits. A tumor growing in the pituitary can affect sleep directly, or cause other medical problems that affect sleep.

Arthritis, Fibromyalgia, and Chronic Fatigue Syndrome

These conditions share several features, such as pain, disturbed sleep, and non-restorative sleep. When patients awaken they may feel as though they have not slept at all or not slept enough.

Arthritis. Many types of arthritis are caused by painful chronic inflammation (or destruction) of the joints. Some affect the larger joints (for example, the hips and the knees), while others affect smaller joints of the hands and feet. These conditions can lead to serious insomnia. Diseases of joints such as rheumatoid arthritis are about three times more common in women than men.

Fibromyalgia. Women in particular are often affected by a condition called fibromyalgia, which is also associated with excessive sensitivity to and perception of pain in the muscles and elsewhere. (This condition is about nine times more common among women than men.) People with fibromyalgia report that the pain is greater following a poor night of sleep. Any painful sensation can lead to trouble sleeping, daytime sleepiness, fatigue, or tiredness. Treatments include pain medicines (but not opiates), antidepressants, and muscle relaxants. The following medications have been approved by the U.S. Food and Drug Administration specifically for fibromyalgia: pregabalin (Lyrica), duloxetine hydrochloride (Cymbalta), and milnacipran HCl (Savella). Some patients might not respond to treatment.

Chronic fatigue syndrome. People with chronic fatigue syndrome (CFS) have as their major complaint overwhelming fatigue (not sleepiness) that does not improve with rest. The disorder has also been called myalgic encephalomyelitis (focusing on the muscle and brain component) and systemic exertion intolerance disease. Patients with CFS cannot tolerate physical exertion; they become severely fatigued for long periods (often more than twenty-four hours) after any exertion. In addition to problems with memory and concentration, these patients share several symptoms with those who suffer from fibromyalgia: pain, insomnia, and nonrestorative sleep. Medical science has not yet determined the cause of this syndrome or found a specific treatment. Although it might seem counterintuitive, a 2016 report found that an exercise program might help the sleep and fatigue symptoms of these patients.

Cancer

Sometimes an undiagnosed cancer might present with sleep symptoms. Recall the patient described in Chapter 6 with restless legs syndrome who was found to have a low level of ferritin, a marker of low iron levels in the body. It turned out that he had a slow-bleeding cancer in his gastrointestinal tract, and his sleep symptoms preceded the diagnosis of his cancer. Night sweats have been seen in some patients with cancer—for example, those with lymphoma.

People with a diagnosed cancer might experience sleep problems for a variety of reasons. The diagnosis itself causes stress, which can lead to insomnia. Additionally, cancer of a particular organ can cause sleep problems related to that organ. For example, a person with lung cancer might awaken with shortness of breath, or a person with cancer affecting the bones may have trouble sleeping because of pain. Finally, some of the treatments for cancer, such as chemotherapy, can result in severe symptoms such as nausea and vomiting, which can also affect sleep.

One 2002 study in the United States reported that the most common problems among cancer patients were fatigue (44 percent of patients), restlessness in the legs (41 percent), insomnia (31 percent), and excessive daytime sleepiness (28 percent). Sleep problems were most prevalent in patients with lung cancer. (A side note: since 1987, more women in the United States have died of lung cancer than of breast cancer.) Insomnia and fatigue were more common among patients with breast cancer.

Some treatments for breast cancer or ovarian cancer can lead to the immediate onset of menopause; we saw in Chapter 5 that this can interfere with sleep. Tamoxifen, for example, a widely used anti-estrogen drug for the treatment of breast cancer, can cause menopausal symptoms, including hot flashes and night sweats that might lead to insomnia. Some cancer patients who have been treated with chemotherapy or radiotherapy might experience severe daytime sleepiness or overwhelming tiredness. Some patients even develop neuropathy (nerve damage) and other symptoms suggesting restless legs syndrome due to chemotherapy drugs.

Many doctors do not ask their cancer patients about sleep problems. In cases in which the cancer is not cured, treatments are available for relief of the pain; help is also possible for patients who have trouble sleeping. For instance, restless legs syndrome can usually be successfully treated. A study reported in 2016 found that cognitive behavioral therapy (discussed in Chapter 19) can improve sleep in breast cancer survivors. If the sleep problem cannot be solved, the patient might want to try a sleeping pill or a wakefulness-promoting medication such as modafinil or armodafinil to allow him or her to stay alert during the day.

Pain and Pain Treatment

Pain from any condition can lead to insomnia. Patients with joint disease might find their sleep disturbed indefinitely unless they find relief from the pain, through either medication or joint replacement surgery. Research has shown that people with painful joint diseases or disc problems of the lower back can develop restless legs syndrome. Sleep is one of the first functions that suffers when people injure their back or sustain other injuries that cause on-going discomfort. The first line of defense is to treat the problem causing the pain. If that does not work, the patient might need pain relief via medication or massage. If even these are ineffectual, the patient should be referred to a pain clinic. Such clinics are available in many large medical centers.

In several U.S. states doctors can prescribe cannabis to treat pain. (Some states have even approved the decriminalization of marijuana used for recreational use.) Several research studies have shown that cannabis can alleviate pain and improve sleep for some people. But it has also been reported that chronic cannabis users may experience very disturbed sleep when trying to stop.

Traumatic Brain Injury

People who have had a traumatic brain injury can be left with severe day-time sleepiness or insomnia. Such an injury can occur as a result of military activities, motor vehicle accidents, or sports injuries. Right after the injury, most patients spend more time in bed and sleep more. This might be re-lated to damage to the brain centers that control sleep and wakefulness. The sleepiness might continue long-term. This symptom has been shown to im-prove with armodafinil, a medication used to treat the sleepiness of narco-lepsy.

Back to the Woman with a Black Curtain

An overnight sleep test of the patient who complained of a black curtain that blocked her creativity confirmed that her breathing pattern was consistent with heart failure. In the sleep lab, she was tested while breathing oxygen supplied by small tubes ending near her nostrils. This improved her breathing and she slept more deeply. She was sent home on oxygen.

Once her breathing improved, her "black curtain" lifted. She was able to muster enough creative energy to put her art show together. But although her sleep was markedly improved, she still had an abnormal heart, which could not be treated further. Doctors can learn an important lesson from this case:

Insomnia has many forms, many symptoms, and many causes. It can alert doctors to serious medical conditions, and by so doing save patients' lives. In some cases, the last thing a person with insomnia needs is a sleeping pill. In every case, the first thing she or he needs is a diagnosis.

Many medical problems cause sleep difficulties, and sometimes diagnosing the sleep difficulty can uncover a medical condition. Because some medical problems are generally associated with men, doctors are more slow to diagnose them in women (and vice versa). It is never a good idea to ignore a sleep problem.

16

Psychiatric Disorders That Affect Sleep

THE MYSTERY. Do sleep problems cause psychiatric disease? Or is it the other way around? Sleep problems are an extremely common feature of all psychiatric disorders, while the medications used to treat psychiatric conditions can cause insomnia, daytime sleepiness, and restless legs.

The Case of the Woman on Stress Leave

When I first see certain patients, I notice that they look intensely sad, and I know that no matter what tests I do or what medical problems I find, I might never get to the root of their problems. Such was the case with a woman in her sixties who had been referred to me because she had severe sleepiness.

I studied her appearance for clues that might shed light on the cause of her problem. She was roughly fifty pounds overweight and poorly groomed; her hair was uncombed and her clothes were ill fitting and old. The expression on her face was sad and withdrawn.

Her doctor suspected that her daytime sleepiness might be caused by sleep apnea, so he sent her to me for evaluation. She did have the common symptoms of sleep apnea. She snored, stopped breathing during sleep, and was overweight. But there was more. Many nights she experienced severe restlessness and difficulty falling asleep.

When I asked her about her work, she explained that she was on medical disability—stress leave. She had been a senior executive at a bank, in charge of the loan department, but was now unable to work. She blamed her current situation on the poor economy and a nervous breakdown; her doctor was treating her for depression. She was also convinced that if her poor nighttime sleep and daytime sleepiness could be treated, all her problems would go away. It turned out that her treatment for severe depression involved several medications that could both disrupt her nighttime sleep and make her sleepy during the day. But she might also have a sleep disorder such as sleep apnea that could worsen her depression. I knew that this was going to be a tough case.

Her sleep test showed that she moved a great deal during sleep and she stopped breathing about six times an hour. Although her breathing pattern improved when she was tested while being treated with continuous positive airway pressure, her sleep was still unstable, partly because of excessive movements. I knew the cause of her sleep-breathing disorder, but the solution for the movements would be more difficult.

Mental Disorders and Sleep

Mental disorders of various severity are very common, and they often coexist with sleep disorders or cause sleep problems. In a survey of European countries published in 2011 it was shown that over a third of the total European population suffers from a mental disorder of some sort. The two most frequent were anxiety disorders (affecting 14 percent of the population) and depression (affecting 6.9 percent of the population). The report also found that insomnia affected 7 percent of the population.

The costs of treating mental disorders are huge. Another European study in 2011 reported the costs to society of mental and sleep disorders. The cost in billions of euros broke down as follows: addiction, 65.7; anxiety disorders, 74.4; mood disorders, 113.4; psychotic disorders, 93.9; and sleep disorders, 35.4.

Symptoms of a mental disorder and sleep problems are often interwoven. Sleep disturbance is a very common occurrence in mental illnesses, and the disturbed sleep, in turn, can cause daytime sleepiness and other symptoms, which can then worsen the symptoms of the mental disorder. In fact, disturbed sleep can itself cause a mental disorder. Furthermore, some of the symptoms of sleep disorders are similar to the symptoms seen in psychiatric conditions, and many people with sleep disorders are misdiagnosed as having a mental condition.

To confuse matters even further, the drugs used to treat psychiatric conditions frequently cause sleep disturbance. Often it is very difficult to figure out the source of a particular symptom. Medications used to treat depression, for example, can cause restless legs syndrome (see Chapter 11), and drugs that treat schizophrenia can cause weight gain, which in turn may cause sleep apnea (Chapter 12).

The category "psychiatric illness" or "mental illness" lumps together many conditions in which patients experience changes in mood, perception of reality, thought processes, or behavior. Traditionally, these were not considered medical disorders but were classified as disorders of the mind. We now know that several of these conditions are caused by biochemical abnormalities involving the brain, and treatment often involves correcting the biochemical abnormalities.

The important association between sleep and mental illness cannot be overstated. A 2016 study found that people with any mental disorder who reported disturbed sleep were more likely to have suicidal thoughts or to plan or attempt suicide.

The mental disorders we most often see at the sleep clinic in association with sleep problems are disorders of mood (such as depression and bipolar disorder), disorders of thought (such as schizophrenia), and anxiety disorders (including panic disorder and posttraumatic stress disorder, which is often seen in military veterans).

Disorders of Mood

Mood disorders are broken into two general types, depression and bipolar disorder. People with depression feel sad most of the time even when there is no

apparent reason for the feeling. In children and teenagers, depression sometimes manifests as irritability instead of sadness. Some people have a condition
called dysthymia in which there is a chronic depressed mood, but it is not
severe enough to be classified as a major depression.

DEPRESSION

A previous diagnosis of depression is particularly common in female patients
who come to the sleep clinic. In my practice, about 21 percent of women
referred for sleep apnea were being treated for depression, compared to only
7 percent of men referred for the same condition. About 8 percent of North
American adults can expect to have severe depression at some time in their
lives. Most of those affected will be women. It is important to note, however,
that many of the patients being treated for depression were not depressed.
They were sleepy, and they had been misdiagnosed.

Links between sleep and depression are very strong. Insomnia is present in
about 75 percent of depressed patients, and daytime sleepiness is present in about
40 percent of young depressed adults. Insomnia and sleepiness affect quality of
life, and have been reported to be a risk factor for suicide. Depression is a very
serious condition, and most depressed people have sleep problems.

Depression in adolescents. In my clinic I have seen many children being
treated for depression who were referred because of insomnia or excessive daytime sleepiness. Before puberty, boys and girls experience about the same rates
of depression. Between the ages of eleven and thirteen, however, there is a dramatic rise in the rate of depression in girls. By age fifteen, girls are twice as likely
as boys to have been depressed. The stresses of adolescence, including physical,
emotional, and hormonal changes, seem to affect girls more. Female high
school students have higher rates of depression, anxiety disorders, and eating
disorders than their male counterparts. The reason for this difference is, at
least in part, related to hormones.

A 2016 study found that disturbed sleep in depressed children may be a
predictor of suicidal behavior. It is important to stress that at times patients
might be misdiagnosed as having depression. Sleepiness, an important symptom of narcolepsy and sleep apnea, has been mischaracterized in some patients as a symptom of depression, for example. Similarly, some teenagers
whose circadian clock has changed so that they are sleepy later at night (see
Chapter 8) and have to be dragged out of bed in the morning have been mis-

diagnosed with depression. Imagine carrying a diagnosis you do not have for the rest of your life!

Depression in women. Sex hormones can cause fluctuations in mood that can lead to depression. These hormonally related fluctuations occur during the menstrual cycle (discussed in Chapter 3), during pregnancy and the time following childbirth (Chapter 4), and in the time immediately before, during, and after menopause (Chapter 5). Some women experience severe mood and physical changes associated with the menstrual cycle. Symptoms include irritability, depressed feelings, and physical changes such as bloating, tender or painful breasts, and cramps. This is premenstrual syndrome (PMS). The symptoms are worse during the days before menstruation starts. When the mood changes are severe, the disorder is called premenstrual dysphoric disorder (PMDD). All these conditions can lead to sleep disruption.

Mood swings are also common in pregnancy, and some pregnant women might become depressed. Women who are trying to become pregnant or who are infertile can be under great stress, but there is no evidence that stress alone leads to depression. Nor is there evidence that having an abortion leads to depression.

The days and weeks that follow giving birth are a high-risk time for women who have had a major psychiatric illness. Some women experience postpartum depression, an extreme mood disorder that requires medical intervention. Such women often have had symptoms or a history of depression before they became pregnant. Motherhood, with all its demands and stresses, increases the risk of depression.

Although menopause is a time when women's hormone levels are changing drastically, it rarely leads to depression. Many menopausal women can have sleep problems, however, because of symptoms such as hot flashes.

Hormonal differences offer a partial explanation of why depression is so much more common in women than men, but some scientists also believe that the greater stresses that many women face are another factor. These stresses can include having major responsibilities both at home and at work, being a single parent, trying to make ends meet financially, and being the main caregiver for children and/or aging or sick parents. Many of my women patients who have had sleep problems related to depression were in the process of divorce or were in a poor relationship. Others had children who had marital or other problems. Rates of depression are highest among men and women who are separated or divorced and lowest among those who are married. The quality and stability of a marriage can play a role in depression. Sometimes depression in women is related to a lack of intimacy and a confiding relationship; sometimes it is linked to frequent or severe marital disputes. Women in unhappy marriages have very high rates of depression.

Depression in men. Although depression is less common among men than women, the sheer number of cases still places depression in men as a public health problem. Is there a link between depression and sleep problems? A 2016 study of men found that those with severe restless legs syndrome were very likely to have depression. One important study followed students graduating from the Johns Hopkins University School of Medicine (classes 1948–1964), all men, for up to forty-five years. This study found that those who had insomnia during school were much more likely to develop depression starting about fifteen years after graduation.

I have seen many male patients with insomnia who are victims of the economic recession that started in 2008. Many lost their jobs, or their businesses went bankrupt. Often their minds race when they are trying to sleep because they are worried about their financial situation.

Depression and aging. Depression is not a normal part of aging. Most older people lead satisfying lives and are not depressed. Though there is a perception that they suffer from empty-nest syndrome when their children leave home, research has not confirmed that this situation is a likely indicator of depression.

Sleep problems related to depression. People with depression suffer from a variety of sleep problems. More than half have insomnia; others have trouble staying asleep, or they wake up early in the morning and have trouble falling back to sleep. Because they might be very sleepy during the day, they might take a long nap or drink excessive amounts of caffeinated beverages, both of which could inhibit their nighttime sleep.

Research from Germany published in 2011 showed that people with insomnia were twice as likely to develop depression as people with no sleep complaints. Research from France reported that same year showed that insomnia symptoms, daytime sleepiness, and the use of sleeping pills can each increase the risk of developing depression in the elderly. Research from Japan published in 2010 reported that difficulty falling asleep (but not being able to stay asleep or waking up very early in the morning) could predict the development of depression. Thus sleep disturbance can lead to depression, or be a marker that depression may occur.

About one in five adults in the United States who complain of insomnia are diagnosed with depression. In some people depression manifests as oversleeping. About one-third of all people who experience insomnia will have depression sometime in their lives, as will a quarter of those who have daytime sleepiness. Half of patients with both symptoms will have depression at some point in their lives. Thus, there is a very strong correlation between depression and poor sleep. In fact, one of the symptoms used by the National Institute of Mental Health in the diagnosis of depression is insomnia or excessive sleepiness.

I have seen many cases in which stress in the workplace was the cause of severe insomnia. One of my patients was an air traffic controller who was no longer able to deal with the stress of her job. She became clinically depressed, could not sleep, and nodded off while working. She had to go on disability.

Symptoms of depression. According to guidelines published by the National Institute of Mental Health, if three to five or more of the following symptoms are present for more than two weeks, they may indicate depression.

- Persistent sad, anxious, or "empty" feelings
- Feelings of hopelessness or pessimism
- Feelings of guilt, worthlessness, or helplessness
- Irritability or restlessness
- Loss of interest in activities or hobbies that were once pleasurable, including sex
- Fatigue and decreased energy
- Difficulty concentrating, remembering details, and making decisions
- Insomnia, early morning wakefulness, or excessive sleeping
- Overeating or appetite loss
- Thoughts of suicide or suicide attempts
- Aches or pains, headaches, cramps, or digestive problems that do not ease even with treatment

Treatment of sleep problems related to depression. I am not a psychiatrist. I see hundreds of people a year who have sleep problems who are also depressed, and if I believe that their sleep problems are caused by depression

I recommend they see a psychiatrist or a psychologist. Most medical practitioners use medications to treat depression. More than twenty-five different types of antidepressants are available, and between 60 and 80 percent of people with depression respond favorably to one or more of these medications. It may take several weeks or months of treatment before progress becomes apparent. Sometimes the sleep problem will resolve itself before the mood improves and sometimes after.

Doctors who treat depressive patients with antidepressants might also add a hypnotic medication in the form of a sleeping pill (see Chapter 20). Some antidepressants produce side effects that affect sleep, such as insomnia, excessive sleepiness, or difficulty falling asleep because of symptoms of restless legs syndrome. A patient who is prescribed an antidepressant that keeps him or her awake might be advised to take the medication in the morning. Similarly, if a person is given an antidepressant whose side effect is to make the user sleepy, it might be better to take the medication at night. After evaluating a patient who has been referred to me for insomnia, I sometimes recommend that the doctor prescribing the medications reconsider the drugs being prescribed or change the timing of when the patient takes the medication.

Antidepressant medications. The older antidepressant medications are the tricyclics (examples include amitriptyline and imipramine) and monoamine oxidase inhibitors (MAOIs; examples include phenelzine). These act by increasing the levels of neurotransmitters in the brain (mainly serotonin and norepinephrine), which improves communication between brain cells. It is believed that abnormal levels of neurotransmitters contribute to depression.

Newer medications that primarily affect serotonin levels in the brain and are thought to improve levels of neurotransmitters are selective serotonin reuptake inhibitors (SSRIs). These include fluoxetine (Prozac), paroxetine (Paxil), sertraline (Zoloft), fluvoxamine (Luvox), citalopram (Celexa), and escitalopram (Lexapro).

Another newer class of antidepressants that are thought to improve levels of neurotransmitters are the serotonin and norepinephrine reuptake inhibitors (SNRIs), which include venlafaxine (Effexor) and duloxetine (Cymbalta).

Other newer drugs that affect levels of brain chemicals are bupropion (Wellbutrin), trazadone (Deseryl), nefazodone (Serzone), and mitrazipine (Remeron).

Generally speaking, the newer medications have different—but fewer—

side effects from the older medications. Drugs that affect serotonin levels, for example, may cause sexual dysfunction (decreased interest in sex, decreased bodily response during sex, and decreased ability to orgasm), whereas with MAOIs there might also be a drop in blood pressure, weakness, dizziness, or weight gain.

Before taking these drugs, patients should keep in mind that several of the antidepressants I have listed (especially the newer ones) can have insomnia or daytime sleepiness as a side effect. Some might have weight gain (which can cause sleep apnea) as a side effect. Patients being treated for depression who experience major sleep problems should review the benefits and drawbacks of their medications with their doctor.

Herbal products. Although few doctors prescribe herbal medications, public interest has grown for treating both depression and anxiety with natural products. One plant that is widely used in some European countries to treat depression is Saint John's wort (*Hypericum perforatum*). A 2015 analysis of studies of its use concluded that there is some benefit, but the effect appears to be modest compared to a placebo. The U.S. Food and Drug Administration advised the public that Saint John's wort appears to affect the way the body handles certain drugs, including some used to treat AIDS, and could reduce the effectiveness of certain oral contraceptives and anticoagulants. Alternative treatments, though "natural," are not always safe. Patients should tell their doctor if they are using a herbal treatment for depression.

Other types of therapy. Many patients do not want to see a psychiatrist, and they prefer not to take pills. Several types of psychotherapy are used to treat depression without medication. In these treatments, people talk with a therapist to understand and solve (or at least cope with) problems that might have an impact on their depression. I frequently refer such patients for cognitive behavioral therapy (see Chapter 19). Behavioral therapists, who are often psychologists, help people unlearn the behavioral and thought patterns that could be causing or aggravating their depression. Such treatments can also prevent insomnia from becoming chronic. With cognitive behavioral therapy, patients learn to change negative attitudes and behaviors that contribute to or maintain depression. Doctors might also combine psychotherapy with medications.

While most depression responds to medication, if this does not work, for some severely depressed patients electroconvulsive treatment can be highly effective. In this treatment, doctors induce a seizure while the patient is anesthetized. Why this treatment works is totally unknown. While it has been maligned, it can be lifesaving in a severely depressed patient. The main side effects are related to memory, mild headaches, and muscle pains. Another treatment currently being investigated is a device that stimulates the brain with magnetic waves. The jury is still out as of late 2016 on the usefulness of this treatment.

Because depression is so common and along with its treatment so frequently causes sleep problems, it is important that people being treated for depression make sure the doctor is aware of any sleep difficulties they already have or develop after starting a treatment.

BIPOLAR DISORDER

People with bipolar disorder, also called manic depressive disorder, have features of both depression and mania. When the patient is in a depressive phase, the symptoms are similar to those of depression. During a manic phase, however, the patient might feel inappropriately elated and appear to have boundless energy; patients in this phase might also do inappropriate things such as going on spending sprees. Bipolar disorder has been classified as having two types:

Bipolar I Disorder: defined by manic or mixed manic depressive episodes or severe manic symptoms

Bipolar II Disorder: defined by a pattern of depressive episodes and hypomanic episodes, but no full-blown manic or mixed episodes.

Bipolar disorder often causes severe sleep problems when people are in their manic phase. Some scientists theorize that an abnormal circadian system might play a role. Recent research shows that variations in the genes that control the circadian clock might affect the rate of suicide in these patients. Patients with bipolar disease sometimes have a great deal of difficulty falling asleep at night, and their days and nights might switch. People might report that they do not need as much sleep as they did and that they feel terrific after only two or three hours of sleep. When people switch from the depressed phase to the manic phase (sometimes even before the switch), they experience several days of very poor sleep. Some patients claim they do not sleep at all during this time. Additionally, in what has been called a mixed state of bipolar disorder, patients have mania and depression simultaneously. Almost all these patients have a sleep problem, because it is likely that the chemical changes that are affecting their mood are also affecting the parts of the brain controlling sleep.

Symptoms of mania. The mania symptoms commonly found in patients with bipolar disorder listed below were adapted from the National Institute of Mental Health guidelines.

Mood Changes

Feeling very "up," "high," or elated
Having a lot of energy
Feeling "jumpy" or "wired"
Being agitated, irritable, or "touchy"

Behavioral Changes

Having increased activity levels
Having trouble sleeping
Becoming more active than usual
Talking very fast about many different things
Feeling as if their thoughts are going very quickly
Thinking they can do a lot of things at once
Doing risky things, such as spending a lot of money or having reckless
   sex

Treatment of sleep problems in bipolar disorder. Doctors often prescribe medications to treat bipolar disorder. The treatments, however, might make patients sleepy—very sleepy—and might have potentially severe side effects. When patients have a severe manic episode (in Bipolar I) they could require hospitalization.

Lithium carbonate, a mood stabilizer, is the most widely used medication to treat bipolar disorder. If depression is also a major problem, the doctor might also prescribe an antidepressant. Treatment with lithium has two important effects on sleep that patients need to keep in mind. First, lithium may lead to an underactive thyroid gland, which could cause sleepiness, weight gain, and even sleep apnea. Second, some studies have suggested that lithium can slow the body clock, setting patients on a night-owl schedule (see Chapter 8).

Aripiprazole (Abilify), an antipsychotic agent, has been approved by the FDA for treatment of several mental disorders, including schizophrenia and bipolar disorder, and as an additional treatment for major depression when other antidepressants are not effective.

Valproic acid or divalproex sodium (Depakote) is generally as effective as lithium for treating bipolar disorder. Lamotrigine (Lamictal), another anti-epilepsy (anticonvulsant) drug, is also used to treat bipolar disorder. Other anticonvulsants are sometimes prescribed for treating bipolar disorder, including topiramate (Topamax), gabapentin (Neurontin), and oxcarbazepine (Trileptal).

Valproic acid, lamotrigine, gabapentin, and several other anticonvulsant medications, as well as aripiprazole, can increase the risk of suicidal thoughts and behaviors; people taking these drugs should be closely monitored for symptoms of depression, suicidal thoughts or behavior, or unusual changes in mood or behavior. The FDA has warned people taking these medications to not make any changes in their treatment without talking to their doctor. All the anticonvulsant drugs can make the patient sleepy, which can be a help at night but is less helpful if the sleepiness persists after the patient awakens.

A SAD TALE (OR IS IT SSAD?)

In 1976 Norman Rosenthal moved from South Africa to the United States to train in psychiatry. He noticed in himself that during the winter, especially in the morning darkness, he felt his energy level plummet; when spring arrived and the mornings brightened, his energy level soared. After research and collaboration with scientists working on the circadian system in 1984, he published the first article on seasonal affective disorder (SAD). Just as sleep apnea was not a new disease when it was first diagnosed in the mid-1960s, SAD did not suddenly appear in the 1980s. That the seasons can have a major effect on the body has been known for centuries. (The body of a hibernating bear, for example, changes dramatically before the winter.)

Rosenthal described a group of patients with SAD, most of whom had bipolar disease, whose depression worsened during the winter months (starting in October–December in the Northern Hemisphere). Their symptoms were severe daytime sleepiness, excessive eating, and a craving for carbohydrates. These symptoms improved as the days lengthened in the spring.

For some patients, the symptoms of depression worsen or appear only when nights are longer (during the winter), which is why the condition has been called winter depression. SAD is thought to be not a separate condition but a variant of clinical depression. A milder form has been called subsyndromal seasonal affective disorder (SSAD).

The number of people with SAD increases in areas closer to the North and South Poles, where there is the least amount of daylight in the winter. More cases of SAD have been reported in Alaska (almost 10 percent of the population) than Florida (less than 2 percent of the population). Women are more likely to have SAD than men.

What causes SAD? It is not simply the amount of daylight exposure. Recent research suggests that susceptibility to SAD could be related to variations in genes responsible for the production of a pigment called melanopsin. Specialized cells in the eye contain the pigment, which is light sensitive, and are believed to play a role in resetting the body's circadian clock. This may explain why some northern populations (for example, people of Icelandic descent in Iceland and Canada) are less likely to develop SAD than southern populations.

Why do patients with SAD gain weight? Some researchers have introduced the phrase "circadian desynchrony" to describe the mismatch between an individual's circadian clock and the world he or she lives in. Patients with SAD spend more time in bed awake; it takes them more time to fall asleep; and they have poor sleep quality. The combination of factors can affect several hormone systems, including those involved in appetite control and metabolism. People with SAD often gain weight during the winter months. Reduced sleep time changes the levels of hormones that control appetite (leptin and ghrelin) and can negatively impact the way the cells in the body respond to the hormone insulin. These changes can lead to weight gain.

Treatments have been developed for SAD patients that focus on exposing them to longer periods of bright light. Starting in October and throughout the winter season (in the Northern Hemisphere) a thirty-minute exposure to bright light (natural sunlight or a "light box") can be effective in treating or preventing SAD. A lower intensity and duration of exposure to blue-enriched light has also been effective. Antidepressants might be prescribed when depressive symptoms are present. Some doctors prescribe the hormone melatonin. A new antidepressant compound, agomelatine, which is available in some countries though not the United States, might be effective. This medication stimulates melatonin receptors in the brain and antagonizes serotonin receptors. No scientific evidence exists at this writing that either melatonin or agomelatine has a positive effect on SAD.

SAD can be a serious disorder. Some sufferers have had thoughts of suicide. People exhibiting significant symptoms of SAD must seek professional help. It has been reported that light therapy can lessen suicidal thoughts in patients with SAD. People with SAD should consult a clinician who is familiar with all aspects of treating SAD and should follow the treatment recommendations of their clinician. People with SAD should not treat themselves.

Disorders of Thought: Schizophrenia

Schizophrenia is a devastating illness affecting roughly 1 percent of the world's population. A high suicide rate is associated with schizophrenia, and in spite of marked improvements in treatment, about 20 percent of affected people are incapacitated by the illness. Those who do not require intensive treatment often require many visits to their health care providers and might need hospitalization.

Causes and symptoms of schizophrenia. People with schizophrenia have problems in how their brain deals with thoughts, problems with the content of the thoughts themselves (delusions), and the belief that their delusions are real. In other words, their thinking process becomes illogical, disorganized, and sometimes repetitive. They might experience either delusions (false, unchangeable, and irrational beliefs) or hallucinations (sensations, sounds, sights, touches, tastes, and smells that are not present). Hearing voices that other people do not hear is the most common type of hallucination in schizophrenia, as was depicted in the movie *A Beautiful Mind.* People with schizophrenia might start to believe that they are being followed, persecuted, robbed, or poisoned. They might develop bizarre behavior or ignore personal hygiene.

People with schizophrenia have extremely abnormal sleep patterns. Their dreams can be terrifying, and it can take them hours to fall asleep. They are also likely to have nightmares. I will always remember the videotape of one patient we saw who had awakened during the night with a terrible hallucination, delusion, or nightmare; he started hitting his own head to try to stop the terrible thoughts. Several of the patients I have seen have become nocturnal: they are awake at night and asleep during the daytime, and they come to the clinic complaining of insomnia. Though the sleep problems in people with schizo-phrenia are severe, they can often be treated.

Treatment of sleep problems in schizophrenia. As with the other mental disorders, the treatment focuses on the underlying schizophrenia problem. There are excellent medications specific for this condition now available that help control the disease. Almost all schizophrenic patients I've seen in the sleep disorders center were already being treated for their schizophrenia but still had sleep problems. They were convinced that if their sleep problem could be solved the schizophrenia would go away. Unfortunately, this is not the case. Though some medications used to treat schizophrenia also help to improve the patient's sleep, the sleep does not usually become completely normal. But because it can improve sleep, doctors sometimes prescribe the medication to be taken at bedtime.

A large number of schizophrenic patients have other sleep problems as well, such as obstructive sleep apnea or a movement disorder. It is estimated that over 50 percent of outpatients with schizophrenia are at high risk of hav-ing obstructive sleep apnea. This might in part be related to the fact that some medications used to treat schizophrenia cause weight gain, and this increases the risk of having sleep apnea. Since the medications also cause sleepiness, doctors might not consider whether the patient has developed sleep apnea, so patients or their families should describe the symptoms carefully. Schizo-phrenic patients who develop obstructive sleep apnea are treated in the same way other patients with the disorder are treated.

We have also had patients at the sleep clinic who had narcolepsy but were misdiagnosed as having schizophrenia and treated for this condition instead. One young teenager was even hospitalized in a psychiatric ward. Narcolepsy is a disorder in which people have vivid dreams at sleep onset, but they gen-erally recognize that the images they have are dreams and are not real (see Chapter 13). The schizophrenic patient, when untreated, believes the halluci-nations to be real.

Anxiety Disorders

Tomorrow's the big day: You are about to propose a toast at your best friend's wedding, to be interviewed for a job, to take a final exam, to go onstage in a musical, to take your first overseas plane trip. You are trying to fall asleep but you cannot because you have butterflies in your stomach and your heart is pounding. It is normal to feel nervous or anxious in such stressful situations, and the feeling usually goes away as soon as you stand up for the toast, sit down for the interview or exam, step onto the stage, or fasten your seatbelt. For some people, however, these feelings of fear and dread come on at the wrong time and in the wrong place. Feelings of anxiety take over their lives and prevent them from performing important—or even daily—tasks and activities. These feelings are brought on by an anxiety disorder.

A 2016 study reported that about 16.6 percent of adults worldwide will have an anxiety disorder in their lifetime. These disorders are twice as common in women than men. They include panic disorder, generalized anxiety disorder, social phobia, obsessive-compulsive disorder, and posttraumatic stress disorder. All these disorders can be treated, but they can worsen if they are left untreated. Medications (most often antidepressants, sometimes anti-anxiety drugs), psychotherapy, and cognitive behavioral treatment are the most common treatments. Patients with these disorders, even when they have sleep complaints, are best treated by a psychiatrist. The medications commonly prescribed for anxiety disorders, used at bedtime, could help alleviate the insomnia.

PANIC DISORDER

Your heart is beating fast, you are breathing deeply, you are sweating and shaking, time seems to be standing still, and you feel dizzy. You might have chest pain or tingling in your fingers. You feel as if you are about to die. You have these episodes over and over again, yet the doctor never finds anything abnormal during your checkups. When people have these symptoms for more than a month, they are usually diagnosed with a panic disorder.

Sufferers of panic attacks usually connect them with the situations in which they occur. This can lead to a fear of a particular situation, which the person then tries to avoid. More than half of people with panic disorders awaken with nighttime panic attacks, and many develop a fear of falling asleep. Waking up at night afraid of dying is a terrifying symptom. When such patients come to the sleep clinic, we might screen them for a sleep disorder (patients with sleep apnea might awaken in a panic), but if they don't have a sleep disorder we refer them to a psychiatrist.

GENERALIZED ANXIETY DISORDER

You are always worried about your job, family, or health, even when all seem to be well. You have trouble controlling the worrying. This problem is common in both women and men, and it usually begins to affect people in their early twenties. Most people with generalized anxiety disorder have trouble sleeping because they can't stop worrying at bedtime; added to their other worries will be worry about not falling asleep. Fifty to 75 percent of these patients have difficulties with sleep.

SOCIAL PHOBIA

You are afraid of being embarrassed or humiliated and are uncomfortable in situations that involve social interactions or that might draw attention to you. You experience extreme shyness in meetings, at parties, in the classroom, or even at a restaurant. Students with this phobia might start cutting classes. Other patients might develop panic attacks. Some might start depending on alcohol to relax, and that can result in a new set of problems. About 20 percent of these patients have insomnia.

OBSESSIVE-COMPULSIVE DISORDER (OCD)

This disorder, which affects about 2 percent of the world's population at some time, has two types of symptoms. Obsessions are thoughts or ideas that even the patient will sometimes admit are "crazy," "silly," "pointless," "stupid," or make no sense. In spite of knowing that the thoughts are irrational, the person cannot seem to keep from thinking about them. Compulsions are behaviors that occur in response to the obsessions. For example, the person might believe (have the obsession) that the gas burner has been left on and will keep checking the burner (the compulsion) over and over again. The compulsions are usually repeated each time in exactly the same way. People with OCD do not usually have sleep problems unless they are worrying at night or their obsession leads them to compulsive behavior at night. This might include repeated checking to make sure the doors are locked, the windows are closed, the baby is breathing, or the water faucets are turned off. There are medications that can treat OCD. Antidepressant drugs that affect brain levels of serotonin have been shown to improve symptoms in some patients. Medications approved by the FDA for use in the treatment of OCD include the following antidepressants: clomipramine (Anafranil), fluoxetine (Prozac), fluvoxamine (Luvox), paroxetine (Paxil), and sertraline (Zoloft).

POSTTRAUMATIC STRESS DISORDER (PTSD)

Almost seventy years after being liberated from a concentration camp, a Holocaust survivor still wakes up almost every night with nightmares. A woman is brutally raped and she relives the horror every night in her dreams. About 10 to 30 percent of people who have been involved in a traumatic event develop PTSD.

This disorder, now called posttraumatic stress disorder, has had other names over the years, including "shell shock" and "battle fatigue." People develop PTSD after experiencing or witnessing something horrible and reacting with intense fear, helplessness, or horror. It is common in both the military and civilian populations. PTSD sufferers replay the terrible events frequently, often awakening from nightmares in a sweat with their hearts thumping, sometimes screaming. Some patients exhibit symptoms of panic disorder. Such patients may develop a fear of falling asleep. Many combat veterans have sleep apnea, and many (perhaps most) have PTSD. Some of these veterans will have claustrophobia (due to past experience with gas and other masks) and will thus have difficulty tolerating the CPAP treatment. Some combat veterans even years after discharge from service will awaken at night and "patrol" their dwelling.

Despite its association with the military, PTSD is twice as common among women as men in the general population. When a patient is referred to the sleep clinic for insomnia, sometimes the assessment identifies the cause as a very traumatic event. These patients need psychiatric care. A study published in 2012 showed that a medication called prazosin (a drug originally introduced to treat high blood pressure) along with cognitive behavior therapy (see Chapter 19) were effective for many patients in treating the sleep disturbances in PTSD.

Back to the Woman on Stress Leave

The patient told me that she had put on a great deal of weight, a side effect of one of the medications she was taking because she was severely depressed and had considered suicide. Some antidepressants cause restlessness before and during sleep and increase movements during sleep, which was the problem that was uncovered in her sleep test. Additionally, one of the medications that she was using for anxiety had a side effect of making her extremely sleepy during the day. Because she needed these medications for her psychiatric condition, we could not suggest that she stop using them. I suggested instead that we treat the mild sleep apnea with CPAP. I also recommended that her psychiatrist consider adjusting her medication in the hope of minimizing side effects. The problem was that the sleep disorder was not the cause of her psychiatric disorder. The medications she took for her psychiatric disorder were having profound adverse effects on her sleep. Her sleep was unlikely to improve significantly until the darkness of her depression finally lifted.

Psychiatric conditions often result in sleep problems. Many of these psychiatric conditions are more commonly found in women, so sleep disorders tend to be common in women with these conditions. But addressing the psychiatric problem with medications can exacerbate the patient's sleep problems. Sometimes there is no easy solution to a patient's sleep problems.

17
Medications That Contribute to Sleep Disorders

THE MYSTERY. Many medications have an effect on sleep. Drugs can have both good and bad effects on sleep. Many prescription and over-the-counter preparations, alcohol, and recreational drugs can play havoc with nighttime sleep or cause daytime sleepiness.

The Case of the Executive with Lifelong Insomnia

The patient had come to the sleep clinic because she had insomnia. She was a thirty-five-year-old lawyer working as an executive for a financial company. Her job performance had recently deteriorated, and she was concerned about making mistakes. Her insomnia had begun when she was a child, and her mother had told her she had always been a poor sleeper. She had learned to cope with her difficulty sleeping as a child and later as a student, doing well in school and college.

Then, about six months before she came to see me, her insomnia had worsened to the point where she had asked her doctor for sleeping pills. She had always resisted them in the past because she thought they were addictive. But after she started taking the pills, her insomnia became, if anything, worse than it had been. Her husband had started to complain that her sleeplessness and restlessness were causing him to lose sleep, and their bed was always a mess in the morning, with bedclothes kicked into a tangle.

A doctor's most powerful tool is to ask questions. I was pretty sure I would be able to solve her problem with just a few.

What people consume or take into their bodies may make them sleepy or keep them from sleeping. It is important for us all to be aware of the sleep-related side effects of some of the more common drugs and of alcohol.

Drugs That Affect Sleep

Two-thirds of the U.S. adult population use prescription medications each month, at a total cost of about $300 billion. And according to a U.S. government report from 2015, 9.3 percent of males and 12 percent of females take five or more prescription medications each month. A side effect of many medications is sleep disturbance. There are literally hundreds of widely used medications and products likely to cause sleep problems, so I shall discuss only the most commonly used ones here.

Patients who are taking a medication that they suspect might be causing problems with their sleep can look up the product on the internet; product inserts, for example, can usually be found on the manufacturer's website. Key-words such as *insomnia, restlessness, sleepiness, fatigue, drowsiness,* and *somno-*

lence should help them find other information. If the information is too technical, or if the problem seems to persist, the patient should discuss the issue with his or her doctor.

ANTIDEPRESSANTS

Antidepressants are the most commonly used drugs taken in the United States and by people referred to a sleep clinic. Each month 6 percent of all males and 11.8 percent of all females take antidepressants. Depression is a common cause of sleep disorders. But the antidepressants themselves can cause sleep symptoms, and some patients who have sleep disorders that cause sleepiness are incorrectly diagnosed as being depressed and prescribed antidepressants.

It is hardly surprising that antidepressants can have an effect on sleep: their main action affects certain chemicals in the brain, where sleep is controlled. Not all antidepressants, however, affect sleep in all patients. It is sometimes a challenge for the doctor and the patient to figure out which problems are related to the medication and which to the underlying depression, or whether both are causing the sleep problem.

For some people, certain antidepressants can cause sleepiness as a side effect. (Some doctors consider this a benefit.) But this side effect can be inconsistent from person to person: the same antidepressant can cause sleepiness in one patient and insomnia in others. Some patients might even have both side effects simultaneously—they might become very sleepy yet have difficulty falling asleep. These unwanted effects often lessen as treatment continues, as the patients become used to the medication and as the depression lifts. The doctor might change the timing of when drugs are taken to mitigate the side effects. If an antidepressant causes sleepiness, for example, the patient could start taking the drug at bedtime. If it is stimulating, the patient could try to take it at some time other than bedtime if possible. Patients should discuss their treatment schedule with their doctor.

Other drugs used to treat psychological and psychiatric conditions (see Chapter 16) can also affect sleep, so it is always safest to check with a doctor before taking them.

DRUGS USED FOR CARDIOVASCULAR DISEASE

Because cardiovascular diseases are common, medications used to treat these conditions are among the most widely used by the general population.

Drugs used to treat high blood pressure. Beta blockers (for example, pro-pranolol [Inderal]) are used to treat diseases of the heart (including abnormal heart rhythms) and hypertension. In 2012, 7.3 percent of adult males and 9.1 percent of adult females in the United States took a beta blocker. These med-ications can cause nightmares and insomnia, though not all the drugs in this category have these effects. Some beta blockers not only affect the cardiovas-cular system, they make their way into the central nervous system, where they reduce the production of melatonin. Before taking a beta blocker, patients should consult with their doctor about its effect on sleep. Research published in 2012 showed that taking melatonin improved the disturbed sleep of people taking beta blockers.

Alpha-2 agonists (clonidine, methyldopa) can cause nightmares, insom-nia, and daytime sleepiness. Calcium antagonists and angiotensin-converting enzyme (ACE) inhibitors are used to treat high blood pressure; they very rarely cause problems with sleep. However, for some patients the ACE inhibitors can cause inflammation of the upper parts of the breathing passage, and this can lead to obstructive sleep apnea and severe, almost uncontrollable, coughing, which is likely to keep the patient awake at night. Such drugs usually have names that end with "-pril," and examples include lisinopril, ramapril, and captopril. After the patient has stopped taking these medications, it may take one to four weeks for the symptoms to improve. (In rare cases the symptoms might persist for several months.) Since the patients still need to treat their high blood pressure, they are sometimes switched to angiotensin II receptor blockers, drugs whose names typically end in "-artan." Examples include lo-sartan and valsartan.

Cholesterol-lowering drugs. These drugs, called statins—such as rosuvas-tatin (Crestor), atorvastatin (Lipitor), pravastatin (Pravachol), and simvastatin (Zocor)—are used to lower blood cholesterol levels. These are among the most widely used medications and have few effects on sleep. Some patients have reported insomnia when taking these medications, though it is unknown why the drugs might have this effect.

ANTIHISTAMINES

Histamine is a chemical found in some specialized cells of the body. It is released from these cells during an allergic reaction. Histamine itself then interacts with receptors on other cells. There are two main types of histamine receptors (called histamine-1 and histamine-2) in the body. Histamine-1 receptors are found in the nervous system as well as in cells that are activated in an allergic reaction. The antihistamines that were commonly used until recently blocked these receptors, and their major side effect was to cause sleepiness. Diphenhydramine, an old medication, introduced in the mid-1940s, is used in several over-the-counter sleep medications. Similar medications introduced over twenty years ago include the ingredient triprolidine and azatadine and are usually available in North America in combination with pseudoephedrine. People taking these combined medications may find that they have daytime sleepiness and insomnia.

The second-generation antihistamines do not have this side effect. The most widely used second-generation histamine-1 antagonists, cetirizine (Zyrtec in the United States and the United Kingdom, Reactine in Canada), fexofenadine (Allegra), and loratadine (Claritin), cause few symptoms related to sleep. The absence of these side effects perhaps explains why these types of antihistamines have become much more popular in recent years.

DRUGS USED TO REDUCE STOMACH ACID

Histamine-2 receptors are found in cells of the stomach lining. When activated they can cause overproduction of stomach acid. Histamine-2 receptor antagonists—for example, ranitidine (Zantac)—block this receptor directly and thus reduce the production of acid. Proton pump inhibitors such as omeprazole (Prilosec in the United States, Losec in Canada) and esomeprazole (Nexium) have a different effect on these cells, while reducing the production of acid. These widely used medications usually have little direct effect on sleep. However, iron and vitamin B12 are best absorbed when there is acid in the stomach; patients who use these drugs for a long period of time might experience a reduction in the absorption of iron or B12. This in turn can lead to iron or B12 deficiency, a cause of restless legs syndrome.

NASAL DECONGESTANTS

Pseudoephedrine and phenylpropanolamine are ingredients in medications that treat nasal congestion. These medications can sometimes cause insomnia. Phenylpropanolamine has been withdrawn from the market in both the United States and Canada because of concern that the product can cause stroke in rare instances.

ASTHMA MEDICATIONS

Asthma itself can cause insomnia, particularly if the patient is short of breath, wheezes, or coughs at night, disrupting sleep. Most of the treatments currently available that involve puffers (bronchodilators or steroid medications) have little direct effect on sleep. However, patients who overuse over-the-counter bronchial dilator puffers that contain older medications such as epinephrine could experience difficulty in falling asleep, particularly as these medications are stimulants. Medications containing theophylline can also cause insomnia in asthmatics because this medication is chemically related to caffeine.

DRUGS USED TO TREAT PAIN AND JOINT DISEASES

Nonsteroidal anti-inflammatory drugs (NSAIDs), which are used to relieve the pain of joint diseases, have no known effects on sleep. Examples include ibuprofen (Motrin or Advil), naproxen (Aleve), and celecoxib (Celebrex).

Narcotic painkillers such as codeine and morphine lead to drowsiness, but typically what ensues will be a light sleep with many awakenings. Morphine and drugs with morphine-like effects can suppress breathing. I have seen several patients with sleep apnea brought on by the use of morphine or methadone. The sleep apnea of these patients has been called complex apnea, a form of central apnea. These patients might have sleepiness related to both the medication and the apnea.

DRUGS USED TO TREAT NEUROLOGICAL CONDITIONS

Since the brain controls sleep, drugs used to treat diseases of the nervous system are likely to affect sleep. People receiving such medications should be warned about sleepiness as a possible side effect, and they should not drive if they experience this side effect.

Drugs used to treat Parkinson's disease. As we saw in Chapter 15, Parkinson's disease itself can cause sleep problems. The drugs used to treat the condition usually improve sleep and are used to treat restless legs syndrome. But sometimes the same drugs (those that imitate the effects of a chemical called dopamine) might cause the patient to become sleepy in the daytime. These drugs can also cause patients to develop problems with impulse control, leading to compulsive shopping, pathological gambling, or heightened sexual drive. Patients experiencing these symptoms should consult with their neurologist, who can switch them to other products.

Anti-epilepsy medications. There are a very large number of drugs available to treat epilepsy, and some are also used to treat headaches, fibromyalgia, and restless legs syndrome. Most can cause sleepiness in the daytime. Epilepsy must be treated, and unfortunately the patient might have to tolerate the sleepiness in order to control the seizures.

DRUGS USED TO TREAT CANCER

There are many chemotherapy medications used for different types of cancer. I cannot review them in any detail here, but a few points are worth mentioning. Many chemotherapy medications have nausea and vomiting as side effects that can impact sleep. Some of the drugs can also affect the nervous system, and some users develop restlessness and restless legs syndrome. Many people on chemotherapy experience overwhelming fatigue.

About 80 percent of breast cancer cells are estrogen-receptor positive, which means that estrogen stimulates breast cancer cells to grow. Several medications which counteract this effect on these cancer cells are now widely used as additional treatment for breast cancer after surgery and as a breast cancer preventative. There are three types of such hormonal medications:

SERMs *(selective estrogen-receptor modulators):* tamoxifen (Nolvadex, Valodex, and Istubal), raloxifene (Evista), and toremifene (Fareston);
Aromatase inhibitors: anastrozole (Arimidex), exemestane (Aromasin), and letrozole (Femara); and
ERDs (estrogen-receptor downregulators): fulvestrant (Faslodex).

The anti-estrogen effect of these medications frequently causes menopause symptoms, including hot flashes and night sweats, which can result in sleeplessness. (These problems are discussed in Chapter 5.)

Alcohol

Many people use alcohol as a sleep aid. This is ill advised. Although alcohol does make a person drowsy, when the blood alcohol level drops, it activates the sympathetic nervous system, which wakes the person up, speeds up the heart, and might cause sweating and headaches. People should probably not try to sleep until after the alcohol has disappeared from the body to avoid this effect. Roughly one ounce of alcohol is found in two 12-ounce servings of regular beer (usually about 5 percent alcohol), two 5-ounce servings of wine (usually about 12 percent alcohol), and two 1.5-ounce servings of distilled spirits (usually about 40 percent alcohol; some distilled spirits have much higher concentrations), and it takes the body about an hour to clear an ounce of alcohol.

Another negative side effect of alcohol is that it can make some people snore. For some people, for example those with sleep apnea (see Chapter 12), the sleep-breathing problem becomes much worse after alcohol consumption. Hundreds of women have told me that their husband's snoring is worse after he has had a few drinks. When the snorer wakes up with a hangover or severe headache, it might be because of apnea. Alcohol makes people sleepy, and the combination of the alcohol and sleep deprivation or a medication that normally makes the user sleepy can cause the user to pass out. Alcohol use increases sleepwalking in those who have a tendency toward this type of nocturnal wandering. Alcohol is a nervous system depressant, and it can worsen the symptoms of depression.

Drug and Alcohol Abuse

Not all the possible drug abuse problems can be covered here. I shall address only the most common ones. Most important for people taking drugs or alcohol to keep in mind is that all drugs that affect the brain can affect sleep. These include alcohol, caffeine, prescription drugs, illegal drugs such as cocaine, amphetamines, and many others.

Abuse of prescription drugs. Some prescription drugs can be addictive. This is true of narcotic painkillers and is sometimes the case with pills used to treat anxiety and insomnia. Often a patient is not truly addicted but is psychologically dependent on a drug, such as a sleeping pill. Patients taking these drugs should consult with their doctor about ways to avoid becoming dependent on them. (I discuss how to reduce sleeping pill dependence in Chapter 19.)

Alcohol abuse. Alcoholics have many problems with sleep. They may have trouble falling or staying asleep, have episodes of sleepwalking, and be very sleepy in the daytime. When binge drinking, they might drink, then have a short sleep, then drink again, in an ongoing cycle. Many recovering alcoholics find that their sleep does not return to normal. They may continue to have severe difficulty in falling and staying asleep, and might benefit from seeing a sleep specialist.

Illegal drugs. Almost all illegal drugs have an effect on the nervous system and can lead to sleeplessness or sleepiness. A big change in personality and sleep pattern often signifies either a psychiatric problem or a drug problem. Both of these are beyond the scope of this book, and people using them should consult with a doctor.

Club drugs. Just as alcohol was widely available during Prohibition, illegal drugs that affect the nervous system are widely available today, and are mainly used by teenagers and young adults. These compounds, often categorized as club drugs, are taken at bars and dance parties, such as raves, that can last all night. Parents should be aware of the dangers of such drugs to their children. Over 20 percent of high school students have attended a rave where club drugs are available, according to a study from 2015. The National Institute of Drug Abuse posts useful information about current club drugs, including descriptions of their effects and prevalence among certain age groups, on its website (drugabuse.gov).

Stimulants (uppers) such as MDMA (also called ecstasy) are commonly used by students. In the United States about one in ten high school seniors admits to having used it. More than 3 percent admit to having used it at some point in the previous month. Used to increase stamina, in large doses MDMA can cause body temperature to become so high that it can lead to muscle breakdown as well as kidney and cardiovascular system failure. It can lead to heart attacks, strokes, seizures, and death. MDMA can permanently damage brain cells. Insomnia, which can be severe, is common with this drug, and can continue after the person stops using it. Amphetamines, cocaine, and illegally prescribed Ritalin can all cause insomnia.

Depressants (downers) such as rohypnol (flunitrazepam, a medication related to Valium but not approved in the United States), LSD (lysergic acid diethylamide, a hallucinogen), GHB (gamma hydroxybutyrate, also called sodium oxybate, Xyrem), opiates (for example, heroin), and ketamine (an anesthetic) are other popular club drugs, sometimes used to counter the stimulant effects of uppers. Some of these drugs make people pass out and have been used as so-called date-rape drugs. In my experience Xyrem is not a problem for the latter, because it is hard to get (a highly controlled pharmacy controls it for the entire country) and salt is added to it so that people taking it can recognize it. However, GHB can be manufactured easily, and this illegal product is what finds its way into clubs. Without the added salt this product is tasteless, and a person taking it will fall into a deep sleep within minutes.

Back to the Executive with Lifelong Insomnia

The question that started my patient and me down the right path was "What sleeping pill were you prescribed?" She named the medication. It was not a sleeping pill but an antidepressant that she was taking at bedtime. She did not know that she was taking an antidepressant or that a side effect of the medication was to make her more fidgety at bedtime; in fact, she had developed restless legs syndrome, which was what kept her husband awake at night. Before she began taking the "sleeping pills" she had lain in bed calmly waiting for sleep. She now tossed and turned all night. She continued to take the pills because she assumed that her insomnia was getting worse. She had not made the connection between taking the pills and the increase in her symptoms.

The patient had primary insomnia (insomnia whose cause is still a mystery), which she had had most of her life. We discussed treatment options for her. She felt she had to use sleeping pills some nights in order to function at work the next day. We picked one of the new short-acting sleeping pills. A few weeks later she called me to tell me that her restlessness had vanished a couple of days after she stopped taking the antidepressant. She was getting excellent sleep on the new sleeping pill, with absolutely no side effects. She was taking the pills three or four nights a week and was satisfied.

Drugs can affect sleep, and people planning to use medications, whether produced by pharmaceutical companies or available in health food stores, should inform themselves about what they are taking before they do so. Why are they taking the drug? To cure a disease? Has a doctor diagnosed the problem? Are they receiving the right treatment? What are the side effects of the medication? It is important for people to consult with their doctor about all the effects of a particular drug, including its effect on sleep, before putting it in their bodies.

PART FOUR

Getting Help

18

At the Sleep Clinic

THE MYSTERY. If you have a serious sleep problem, the sleep clinic could save your life. Or at least your marriage. At the sleep clinic, patients of all ages with a variety of sleep problems are diagnosed and tested. The most commonly found disorder, sleep apnea, can be deadly if the patient does not get treatment. But getting the wrong treatment can be just as dangerous, and sleep clinics can uncover a wide range of sleep disorders.

~~~~~~~~~~~~~~~~~~~~~

The Case of the Woman Who Did Not Have Sleep Apnea

I was on my way to work at the sleep clinic one morning and was waiting for the elevator in the lobby when I noticed an obese young woman fast asleep in a chair.

Two hours later I saw her again. Again she was fast asleep, but this time she was sitting in the waiting room. When I finally saw her in my examining room, she was still sleepy. She could barely stay awake as we spoke.

The doctor who had referred her was sure she had sleep apnea: she was always sleepy and nodded off whenever she was inactive, at any time of day. She weighed about 350 pounds. She snored. Her father had sleep apnea and was on a continuous positive airway pressure machine. The patient's doctor did not think that she even needed a sleep test: he thought that she should be started on CPAP immediately.

I would never start someone on CPAP without confirming the diagnosis. Part of the diagnostic process is a detailed interview, and when the patient told me that she had been sexually abused as a child and described the treatment she was receiving for her trauma, I knew that a sleep test was absolutely necessary.

The New Science of Sleep Medicine

In 1970, if you had a sleep disorder, neither you nor your doctor would have recognized it. This is because at that time no one studied—or even knew about—sleep disorders; there were no sleep specialists and no sleep clinics. A few medical centers connected with medical schools had laboratories where researchers interested in what happened to the brain during dreaming or in the relationship between dreaming and mental illness could study aspects of sleep, but that was the extent of it.

The "discovery" of sleep apnea in the mid-1970s changed all that. After doctors and scientists realized that sleep disorders could represent a danger to the sleeper, perhaps even kill him or her, they began focusing on diseases and disorders that could be linked to sleep problems. Between the 1970s and mid-1990s, they carried out extensive research into sleep disorders. A new field of medical science was born.

The university research labs became the nucleus for sleep clinics. Scientists studying sleep formed groups to exchange information, the first medical journal to focus on sleep problems was published, and standards for sleep clinics were established to protect patients. The result of this intensive research was that the medical community suddenly recognized that sleep disorders were very common—even more common than conditions such as asthma that had received much greater attention. Growing awareness of sleep problems, coupled with patient demand, led to more research: government agencies and insurance companies started to fund tests and treatments for sleep disorders. In 1990, the first textbook on sleep disorders targeted toward doctors was published, *Principles and Practice of Sleep Medicine*, which I edited with my colleagues Thomas Roth and William C. Dement. The three of us represented the diversity in the field: Dr. Roth was a psychologist, Dr. Dement a psychiatrist and one of the pioneers in sleep research, and I am a specialist in internal medicine and lung diseases. In fact, I was the only lung specialist at the first sleep meeting I attended in the 1970s; most of the early sleep experts were psychiatrists. I had gone because I was doing research in breathing during sleep, and although I knew a lot about breathing, I knew almost nothing about sleep.

Today thousands of lung disease specialists treat sleep disorders, and the diagnosis and treatment of sleep disorders has become mainstream medicine. Sleep clinics can be found in most major U.S. cities and around the world.

ACCREDITATION

In North America, the quality and the types of sleep medicine laboratory services vary significantly. Whereas, for example, the American Academy of Sleep Medicine (AASM) has been accrediting sleep disorders centers in the United States since 1977, carrying out inspections to ensure that the equipment and staff meet appropriate standards, there is no accrediting organization in Canada. In Europe, accreditation of sleep clinics and specialists varies from country to country, although standards have been recommended by scientific groups. To be accredited in the United States, the staff must be trained and certified competent to examine patients, use the equipment, conduct tests, and analyze the results for the sleep evaluation. The staff must include a board-certified sleep medicine specialist. Clinics with full accreditation must be able to manage all sleep disorders. The AASM also accredits home sleep testing programs and durable medical equipment providers, which supply the machines and disposables used to treat sleep apnea.

In addition to the AASM accreditation, sleep clinics and sleep labs in the United States can choose to be accredited by the Joint Commission. This is the largest U.S. standard-setting and accrediting body for health care, responsible for accrediting all U.S. hospitals. This organization accredits facilities (hospitals, laboratories, homecare programs, and so on) for most types of medical procedures and tests; it does not specialize in sleep accreditation.

Another U.S. agency, the Board of Registered Polysomnographic Technologists (BRPT), is an independent nonprofit organization that since 1978 has certified by examination technologists who work in sleep labs. Technologists have an important role in a sleep laboratory because they are frequently the only people present while the patient is taking an overnight test. If a medical emergency occurs, such as a patient developing a heart arrhythmia, the technologist must be able to recognize and manage it. To be registered, technologists must pass rigorous tests to ensure that they possess the knowledge and skills to conduct overnight sleep studies and to work with patients who have sleep and medical disorders. In some parts of the United States, people who work as technologists must also be registered respiratory therapists.

Starting in 2011 the American Academy of Sleep Medicine has offered certification by examination and BRPT technologists have been eligible to be certified by the AASM. They are designated registered sleep technologists (RSTs).

Why am I giving you so much detail about accreditation? Because in addition to the accredited sleep centers, there are literally thousands of labs and doctors' offices in the United States that offer sleep studies, and some do not meet standards. The clinics and labs vary widely in terms of the type, quality, and cost of the testing. Patients seeking treatment should be aware, however, that an expensive study is no guarantee of quality. About twenty-five years ago, I taught a course for doctors on how to evaluate people with sleep problems. A doctor in the audience asked why we had to measure so many things to make a diagnosis. He had no training in sleep disorders (that was why he had enrolled in the course); for his diagnoses he measured only the level of blood oxygen—and billed patients and insurance companies $1,200 for the test. The equipment that he used for the test cost roughly $2,000, and it was inaccurate. Although he had more than paid off the cost of the equipment in just two days of use, he was still charging double what some of the best labs in the United States charged for a comprehensive study conducted by a trained sleep medicine specialist. I was ashamed at the way this doctor cheated his patients; I was even more ashamed when several other members of the audience expressed an interest in his methods.

TYPES OF SLEEP STUDIES

Sleep tests can be done in a laboratory or in the patient's home. The types of data monitored by these tests can vary: laboratory tests usually monitor ten to sixteen kinds of information, the home tests three or four. The tests available in a given sleep clinic depend entirely on what the local insurance companies will cover. In parts of the United States, for example, managed care companies or insurance companies will not pay for comprehensive evaluations, but often will pay for limited laboratory tests or home sleep tests. Some insurance companies do not provide coverage for patients to see a sleep expert, who is the person best able to determine the appropriate treatment.

Screening studies and home sleep tests. The word *screening* usually means a test—often an initial test intended to identify or catch a potential problem but that might not lead to a diagnosis, such as mammography, which is used to detect breast cancer. Screenings are usually inexpensive and very sensitive. Thus, a problem might be identified that, on further examination, is determined not to be present. But although screening tests can at times be inaccurate, they remain valuable.

Home screening for sleep apnea, in particular, is liable to give a false negative. In 10 to 30 percent of tests that give a negative result, it turns out that the patient does have sleep apnea. Since home tests do not determine whether a person taking the sleep test is, in fact, asleep, the test might register "normal" for a patient who barely slept during the monitoring. Thus when a person with symptoms of sleep apnea tests negative in a home screening, the result can give both patient and doctor a false sense of security. Additionally, since home tests usually screen only for sleep apnea, they will not give information about the many other sleep disorders that could be affecting the patient. These tests generally collect information about the patient's blood oxygen level, breathing pattern, and snoring. Sometimes they include an electrocardiogram. When the home screening test is negative but the doctor strongly suspects the patient has a sleep-breathing disorder, he or she will usually order the test to be repeated in a sleep laboratory to make sure that no serious problem is present.

Patients whose screening test is positive also usually have a second test, in their case to determine the best way to treat the sleep-breathing disorder. If the doctor is fairly sure that obstructive sleep apnea is present and sees no evidence of other problems, the patient will probably be put on a machine to treat apnea without additional testing. Most machines now used to treat sleep apnea can determine whether a patient is using the equipment and whether it is effective. Some U.S. medical insurance companies and Medicare require

confirmation that the patient is using the equipment appropriately and is ben-
efiting from the treatment.

Comprehensive sleep studies. Comprehensive testing done in sleep laboratories measures all the information that is required for a doctor to make a
diagnosis. A few portable systems are even available that can make all these
measurements in a patient's home. Comprehensive systems can check to see
how the person responds to treatment. In addition, sleep disorders often affect
several organ systems, and it is sometimes vital to see whether some of the
measurements indicate that the patient is in dangerous territory. A technologist usually monitors the test in the lab.

When a Sleep Problem Is Suspected

If you think that you have a sleep problem, you need to consult a doctor. Make
sure that you communicate your symptoms clearly (see Chapter 6), and don't
assume that your doctor will ask about symptoms of a sleep problem during a
routine visit for another problem. It's a good idea to write down your concerns
and questions before the appointment so that you don't forget any of them.
Bring a list of all of the prescriptions (including the dosages) and over-the-
counter medications you are taking. If appropriate, bring your bed partner; he
or she might be able to provide important information about snoring, stopped
breathing, movements, and other unusual behaviors you might exhibit while
you sleep.

Your doctor might not be comfortable evaluating or treating these problems and might refer you to a sleep specialist. Do not be concerned if this is
the case. Many doctors have had little or no training in the treatment of sleep
disorders because sleep medicine is still a relatively new field of medicine.

CHOOSING THE DOCTOR AND CLINIC

Every doctor in the United States has a medical diploma, but not every doctor
can manage sleep disorders. Not all doctors have had comprehensive training
in sleep, and only a few have had the specific training and passed the board
examination in sleep medicine. Every doctor can write prescriptions, but not
all doctors know enough about the specialized drugs often used in sleep medicine. You have a right to know whether the person diagnosing and treating you
or a family member has the necessary knowledge and expertise.

Look at the office walls. When you get to the doctor's office, take a look at the walls. Usually the doctor's diplomas are hanging up. See if the clinic is accredited as a sleep disorders center by the American Academy of Sleep Medicine. Has the doctor been certified as a sleep specialist by the American Board of Sleep Medicine or the American Board of Internal Medicine? Look for the word *sleep* on the diploma.

Ask questions. Don't be embarrassed to ask the doctor for his or her qualifications. You might discover that the person sitting in front of you is not a medical doctor. Ask about the doctor's training and experience in sleep medicine.

If you have sleep apnea and are having a sleep test, ask for the name of the doctor who will be interpreting the test, where this doctor is located, and what his or her qualifications are. You might be shocked to learn that your doctor does not know the answer to any of these questions. If this is the case, ask whether the doctor or the clinic is paid separately for doing the sleep test. Some medical offices are paid to offer tests, but their doctors are not qualified to interpret those tests. If you take a home sleep test, it might be interpreted by doctors in another state; their names might not even appear on the test report (and their signatures might be illegible).

Ask the doctor about his or her experience with the various masks and machines that are used to treat apnea. Dozens of masks and several machine types are available. The machines have names like CPAP, AutoPAP, BiPAP, ASV, AutoBiPAP, and AVAPS—if the doctor starts to fumble, be wary about whether he or she has the expertise to treat a condition you might have the rest of your life.

After you have found the right doctor and had a consultation, he or she might recommend you have an overnight test in a sleep lab.

GETTING READY

Some people going to a sleep lab might be concerned about sleeping in a strange environment. If you have special needs or are very modest, let the people in the lab know about your concerns, and they will try to accommodate you. The technologists are professionals, and they have seen it all: four-hundred-pound men with teddy bears, people sleeping in nightshirts and people sleeping in the nude, teenagers sleeping with headphones on, people fanning their feet or pouring water on their legs. If you are nervous, bring a friend or spouse for moral support while the technologists are getting you ready for the test. If your young child is having a sleep study, the staff can probably find you a place to sleep in the lab. The sleep center will give you detailed instructions about the test, what to expect, and what to bring with you (pajamas, toothpaste, and so on). Many labs have showers.

Many people worry that they will not be able to sleep in the strange environment of the lab. We have heard variants of "There is no way I will sleep at all in the lab" and "What will you use to put me to sleep?" hundreds of times. In reality, it is very unusual for a patient not to fall asleep in the sleep lab—even people who complain of severe insomnia. The technologists do not use sleeping gas or sleeping pills to help people fall asleep. Nor do they use needles. The technologists are trained to help patients relax, and they use sensors to gather their information. While the technologist is applying the sensors to the patient, he or she will offer a detailed description of the procedures, explaining, among other things, that during the night the patient might be started on nasal CPAP, a treatment for sleep apnea. The patient will often be shown a video about the condition and about the treatment.

What to Expect When Being Tested in the Sleep Laboratory

At a comprehensive sleep laboratory, the technologist will attach electrodes on your scalp, chin, chest, and legs. He or she will also place sensors that measure oxygen on your earlobe or on your finger and other sensors in front of your nose and mouth to measure when you stop breathing or when you snore. At this point, most patients resemble an alien from outer space. While the technologist is applying the sensors, he or she will usually tell you what is likely to happen during the study, particularly if you are to be treated with a CPAP system. (The patient can also be sent home with such a system.)

The typical sleep apnea patient falls asleep within five to ten minutes and starts to snore and stops breathing fairly quickly while being observed and monitored by the technologist. While the patient is sleeping, the technologist will operate the recording instruments, complex devices that require a highly trained technologist to monitor. Twenty to thirty years ago, most sleep laboratories collected their data on paper. A full night's recording used between six hundred and a thousand sheets of continuous paper. Today all modern laboratories have switched to computerized systems (and thereby saved hundreds of thousands of trees). As with all things electronic, most systems have become smaller, and the equipment used in the sleep laboratory has shrunk from many hundreds or thousands of pounds in weight to as little as one pound. This equipment has devices to measure efforts to breathe (by monitoring the movement of the chest and abdomen), the effectiveness of breathing (by measuring the blood oxygen level), whether breathing stops (by detecting the flow of air in front of the nose and mouth), the heartbeat (using an electrocardiograph, or EKG), and the brain waves used to indicate which stage of sleep the patient is in.

The technologist watches and digitally video records your sleep throughout the study, looking specifically to answer the following questions:

Is the patient asleep? The technologist needs to measure your brain waves using an EEG so that the person analyzing the record will know when you are asleep. Remember that during REM (dreaming) sleep, rapid eye movements occur and the sleeper is paralyzed. Thus, measures of eye movements and muscle tone are required to indicate whether you are in REM or non-REM sleep. It is during REM sleep when the most severe sleep-breathing abnormalities occur. For some people, sleep-breathing abnormalities occur only in REM sleep.

Is the patient breathing? It is important to determine whether you, the patient, are breathing efficiently and to find out the level of effort you require to breathe. These measurements will help determine what type of sleep-breathing problem (obstructive or central apnea; see Chapter 12) is present and therefore what treatment will be the most appropriate.

Does the patient's blood oxygen level change? When a person has a sleep-breathing problem, the blood oxygen level drops when he or she stops breathing. Measurement of the blood oxygen level is critical because severe drops in blood oxygen level bring increased risk of cardiovascular problems such as abnormal heart rhythms.

TESTING ON TREATMENT FOR APNEA

During the night, if the technologist finds that you have a significant sleep-breathing problem, he or she might wake you and tell you that you will now be started on CPAP or other treatment to see whether the treatment solves the problem. This involves placing a mask over your nose and mouth that is connected by a hose to a blower. (This type of study is called a split-night study: the first half is the diagnostic part and the second half is the treatment; the technologist determines how much CPAP pressure you will need.)

The technologist usually has many masks available and will test to find the most comfortable and effective mask. The technologist may change the mask during the night if the mask the patient is using is leaking or not doing the job.

Sometimes the technologist will do a diagnostic study for an entire night and then bring the patient back for further study a second night. Although there are advantages to such an approach, it is more expensive. Some labs have automatic CPAP machines that determine the correct CPAP pressure while the patient is sleeping instead of having a technologist make the adjustments. Sometimes this testing takes place in the patient's home without a technologist. This latter approach might not be as effective as having a trained person present to make the adjustments. The technologist can deal with problems that might come up, such as a patient opening his or her mouth during sleep, which makes CPAP ineffective, or beginning to panic when the CPAP mask is on. Sometimes CPAP turns out to be the wrong treatment. If this is the case, nothing can compare with having an experienced technologist at your bedside.

MEASURING SLEEPINESS AND DETERMINING REM

If you experience daytime sleepiness, you might get another type of test at the sleep lab: a multiple sleep latency test. This test is a measure of your sleepiness during the day and is used to confirm the onset of REM sleep during naps for people with narcolepsy. In this test, you will be given four or five opportunities to nap for twenty minutes every two hours during the day, lying in a quiet, darkened room while hooked up to the same monitors that are used for night studies. Your sleepiness is measured by the amount of time it takes you to fall asleep during these nap opportunities; shorter times indicate higher levels of sleepiness. On average, a person with extreme sleepiness will fall asleep in less than eight minutes during the naps. People with narcolepsy will usually have episodes of REM sleep during two or more of the naps. People without narcolepsy seldom dream during naps.

The Value of a Sleep Test

Sleep testing can find the information that may save your life, your marriage, or your job. The information revealed in a sleep test could result in treatment that might prevent you from having a heart attack or stroke. It could lead to proper treatment of sleepiness so that your child can stay alert and excel in school.

More than eighty sleep disorders have been described, and some can be uncovered only by a comprehensive evaluation and a sleep test. Besides enabling trained personnel to diagnose sleep apnea and adjust CPAP settings, laboratory sleep tests can help reveal narcolepsy, movement disorders, and seizures during sleep. Patients nearing the end of their life, such as people with heart failure, could benefit from the treatment of specific problems found in the sleep test. If you think you have a sleep problem, a sleep clinic could give you the most important night's sleep you'll ever have.

Back to the Woman Who Did Not Have Sleep Apnea

The patient's sleep test did not confirm sleep apnea or even the mild form of apnea called upper airway resistance syndrome that is common among women. My patient fell asleep quickly during the test, and although she snored, her breathing pattern remained normal. She had no REM sleep and very little stage 3 or 4 sleep, but her EEG frequently showed a type of wave that is common in people using certain classes of medications. The sleep test did not show a sleep-breathing problem.

What was causing her sleepiness? Medications. She was on four drugs to treat the posttraumatic stress disorder that had come on as a result of the sexual abuse. One of the drugs often causes weight gain. Three of them cause sleepiness. No wonder she was barely able to stay awake. My recommendation to her family doctor was to refer her to a psychiatrist so that her drugs could be modified. Although it did not confirm sleep apnea, her sleep test led to an accurate diagnosis.

Sleep tests are as vital to the practice of sleep medicine as interviewing and examining the patient. From sleep apnea to stroke, narcolepsy to psychological trauma, the sleep lab helps doctors uncover—and treat—the disorders that affect your life.

19

Beating Insomnia Without Pills

THE MYSTERY. If insomnia has many causes, can it also have many cures? Can it be treated without pills? Many patients can help cure their insomnia without using medications. Behavior modifications can often do more than hypnotics to give a patient a good night's sleep.

The Case of the Woman Who Was Afraid to Make a Fool of Herself

After a referral from her doctor, a twenty-five-year-old woman came to my office and described her insomnia. She was very concerned about it—she even worried that her embarrassing sleep problem might break up her relationship with her boyfriend. When I asked her how insomnia could possibly be embarrassing, she explained that the insomnia was not really the problem. What was embarrassing was that when she did fall asleep, she apparently would get up and start screaming. She feared that her sleepwalking and screaming would drive her boyfriend away, and because she was so worried about this happening, she now had great difficulty falling asleep.

The patient had a lifelong history of sleepwalking and sleep terrors. While she was sleeping she would sometimes get up and walk around; at other times she would yell and scream. She never had any memory of these episodes, but other people would tell her about them, and she found this very embarrassing. She had developed a fear of falling asleep that was closely linked to her fear of embarrassment, and this had created a vicious cycle in which her fear caused sleep deprivation, which in turn exacerbated her episodes of sleepwalking and sleep terrors.

She was becoming desperate, but she made it clear that she did not want to take drugs to help her sleep. I thought there might be another solution.

An Alternative to Pills: Cognitive Behavioral Therapy

People with insomnia can often be effectively treated without pills. Before trying a medication-free approach, however, patients need to be evaluated by a doctor or sleep specialist. As we saw in Chapter 10, insomnia is a symptom and a disorder, and it can have many causes, including medical disorders, psychiatric disorders, and the effects of drugs (including caffeine). Most sleep specialists now recommend cognitive behavioral therapy as the first treatment of insomnia, especially for patients who want to avoid medications.

Cognitive behavioral therapy (CBT) is a psychological approach that includes techniques to treat people with insomnia as well as those with serious depression and anxiety problems. It is provided by many clinical psychologists and other types of health professionals who have had training in this area.

Because it seeks to resolve the problem with behavioral change, cognitive behavioral therapy can help to reduce sleeping pill use. With this approach, the patient receives information about how much sleep he or she needs, learns how to self-monitor sleep, and is encouraged to practice good sleep hygiene. Typically the patient learns to identify and alter behavioral cues that could be promoting sleeplessness and to develop new habits that promote better sleep.

Once patients become more aware of their thoughts and expectations about sleep they might be able to modify them. They also learn different relaxation techniques that allow them to quiet their mind, fall asleep more easily, and have more restful sleep.

Cognitive behavioral therapy is provided by clinical psychologists and by other types of health professionals who have had specialized training in this area. Patients who want to try CBT should check to see if a clinical psychologist is available in their community and whether their health insurance covers such treatment; it often does. Psychologists treat people either individually or in a group setting. Although some people are initially skeptical about the effectiveness of CBT for insomnia, many become strong advocates of this approach once they have followed a course of such treatment (and even begin helping others with their sleep problems!). On average, 70 to 80 percent of individuals experience a significant improvement in their sleep after receiving cognitive behavioral therapy.

The degree of improvement usually depends on how much effort individuals have put forth, so I must stress the importance of being committed to practicing at home. It is important to keep in mind that to see an effect, patients must apply these strategies every night for at least three to four weeks. Practicing these techniques just once or twice, or every now and then, is unlikely to make sustained improvement in the patient's sleep.

Recently, online or computerized cognitive behavioral therapy programs for insomnia have been developed, such as the RESTORE program available through Cobalt Therapeutics (cobalttx.com) in the United States, and the Online Program for Insomnia available through the University of Manitoba (http://www.return2sleep.com/) in Canada, and Sleepio (sleepio.com) in the United Kingdom and the United States. These programs allow individuals to receive help for their insomnia in the comfort of their own homes. However, because individuals have different needs, some people require more sessions than others. Whereas many patients can learn CBT techniques on their own and are successful at improving their sleep, others might require help from a health care professional. Besides the online programs, mobile apps are available for both iOS and Android. These mobile apps are not designed to replace therapists but to supplement the therapy and as homework. Such smartphone apps are being developed continuously.

A Healthy Night's Sleep

A significant problem for many people who experience insomnia is a tendency to become worried and apprehensive about sleep loss. It might help these people to know that research shows that each person has different sleep needs. This is not surprising: people have different shoe sizes, blood pressure, weight—why should their sleep be any less variable and individual? Some people are by nature short sleepers and require only five to six hours per night to feel rested. Other people need nine to twelve hours. Most people fall in between these extremes.

Some of the people whom sleep specialists see feel certain that they need eight hours of sleep a night to function well and to be healthy. These people are often surprised (and relieved) to learn that they can sleep for six or seven hours and feel well the following day. Many people have pressured themselves to obtain an amount of sleep that their body resists. In short, a person who feels rested and alert after fewer than eight hours of sleep a night does not have a sleep problem—that person simply does not need eight hours of sleep a night. Furthermore, it is not a good idea for such people to try to sleep longer than they need. Oversleeping can cause morning headaches, grogginess, and (ironically) sleepiness the following day. Cognitive behavioral therapists suggest that patients conduct individual experiments to determine how much sleep they need to feel rested.

SELF-MONITORING OF SLEEP

A good way for patients to monitor how much sleep they are getting is to keep a sleep diary, such as the one shown in the figure below. Patients should take three to four minutes to complete these diaries before getting out of bed each morning. The diary helps patients collect information about their sleep schedule: how regular their bedtime is, how long it takes them to fall asleep, how many times they awaken in the night, and what time they wake up in the morning. Several devices based on smartphones and "wearables" are used to monitor sleep, but most have not been rigorously evaluated.

When working with a sleep diary, it's important that the patient does not look at the bedroom clock during the night since this can lead to anxiety about not sleeping. Instead patients should guess or estimate how long it took them to fall asleep and how many times they awoke during the night. The sleep diary is an essential part of the therapy because it shows patients the

Name _____   Date _____   Weekday ____ or Weekend ____

Time you turned off lights to go to sleep _____   Time you think you fell asleep _____

It took me _____ minutes to fall asleep last night
I awakened _____ times last night

Time you awakened   Minutes to fall asleep again

_____ am   _____ minutes
_____ am   _____ minutes
_____ am   _____ minutes

I woke up for the last time at _____ am and slept a total of _____ hours

Circle your level of physical arousal when you went to bed
Extremely calm/relaxed   Extremely tense/aroused
                1   2   3   4   5   6   7   8

Write down what you were thinking about as you were in bed: _____
_____

Write down activities from dinner to bedtime: _____
_____

Write down activities once you were in bed: _____
_____

Time you napped   Length of nap   Time of medication/alcohol   Type
_____   _____minutes   _____   _____
_____   _____minutes   _____   _____

Cognitive Behavioral Therapy Sleep Diary

associations between their behavior and their sleep patterns. Someone who is involved in a lot of pre-bedtime activity (for example, doing chores) and is also experiencing a delayed sleep onset might try to schedule a period of time before bed when he or she does nothing but relax. Some patients find that their bedtime varies widely from day to day; such patients could try to establish a more regular bedtime. By opening sleep sufferers' eyes to their sleep behavior, self-monitoring might help those patients close them.

SLEEP HYGIENE

Sleep hygiene refers to habits that promote or prevent sleep. Some of these habits relate to diet, exercise, alcohol and drug use, noise, light, and temperature. In cognitive behavioral therapy, patients will learn to assess their level of sleep hygiene and determine whether modifications are necessary. Some sleep hygiene areas were covered in the "Thirteen Commandments for Fighting Insomnia" (Chapter 10).

One area of sleep hygiene that people with insomnia often need to improve is sleeplessness caused by a snoring or restless bed partner. Cognitive behavioral therapists often encourage couples to consider sleeping in separate beds or even separate rooms to evaluate the impact of separation on their sleep. Though they might be concerned about the effect this will have on their relationship, couples usually find that their sleep improves, and they discover other ways to be intimate without sleeping in the same bed.

Another problematic area of sleep hygiene is sleeping in the same room as the family pet. Dogs and cats can disturb sleep by yawning, snorting, gasping, or moving around. Some patients feel guilty about shutting their pets out of the bedroom, but their sleep can dramatically improve if they do so. They can show their care for the pet by preparing a special place for it with a nice blanket and favorite toy outside of the bedroom.

Cognitive behavioral therapists will also ask patients about their typical exercise pattern because it is related to their adherence to good sleep hygiene. Many people juggling multiple roles of caring for family and working outside the home find they have little or no time for exercise during the day. If they try to make up for this lack by exercising at night or within three hours of bedtime, they might find themselves experiencing sleep problems. Although exercise can leave them feeling tired and relaxed immediately afterward, many people experience an energy burst later that interferes with their sleep. People who work outside the home might want to consider joining a health club near the workplace where they can work out during the lunch hour.

No discussion of sleep hygiene can be complete without mention of the negative effects of caffeine (typically found in coffee, tea, colas, and chocolate), alcohol, and nicotine (see Chapter 20 for a fuller discussion). Although many people are aware of the impact of caffeine on sleep, some are surprised to learn that even one glass of whiskey or wine can interfere with their sleep. Similarly, although some people tolerate caffeine well, many notice an improvement in their sleep when they stop consuming caffeine, even if they have been in the habit of drinking only one or two cups of coffee, tea, or cola a day. Certain individuals appear to be very sensitive to the effect of caffeine. Additionally, nicotine can affect sleep. Smokers who cannot quit could reduce the impact of nicotine on their sleep if they were able to establish regular smoke breaks and smoked the last cigarette of the day several hours before bedtime.

How Cognitive Therapy Works

AUTOMATIC THOUGHTS AND COPING THOUGHTS

Cognitive therapy is based on the idea that our thoughts about various events, activities, and people can affect our feelings and our behavior. Some people are surprised to learn that their thoughts about sleep can affect their sleep behavior. In the context of insomnia treatment, cognitive therapy is used to help individuals become aware of what they are saying to themselves about sleep (their assumptions and beliefs) and to evaluate whether these are realistic and reassuring ways of viewing their insomnia. Therapists train patients to identify "automatic thoughts" (the catastrophizing that occurs when they are having trouble sleeping) and work to replace them with "coping thoughts" (realistic readjustments that recognize that things are never as bad as the insomniac imagines).

Here are some of the more common automatic thoughts connected with insomnia, followed by the coping thoughts that can be used to counteract it and promote better sleep.

> *Automatic thought:* "If I don't sleep well tonight, I won't be able to function tomorrow."
> *Coping thought:* "If I don't sleep well tonight, I'll probably be grouchy tomorrow, but I'll manage."

> *Automatic thought:* "My insomnia is never going to get better."
> *Coping thought:* "My insomnia is a problem now, but if I do the treatment techniques that worked for others, it will get better."

> *Automatic thought:* "I have no control over my sleep."
> *Coping thought:* "My body will tell me when it needs sleep."

> *Automatic thought:* "If I sleep poorly tonight, it will disturb my sleep for the next week."
> *Coping thought:* "If I sleep poorly tonight, I'll sleep better another night."

> *Automatic thought:* "If I don't fall asleep soon, I'm going to be up all night."

Coping thought: "I may not fall asleep soon, but I'll fall asleep eventually. The best thing for me to do is get up and relax until I become drowsy before returning to bed."

Automatic thought: "If I don't sleep well tonight, I'm not going to be able to go out after work, spend time with my family, or do my hobbies."
Coping thought: "If I don't sleep well tonight, I'll probably be more tired tomorrow, but I can still do all the things I want to do even if I am tired."

Automatic thought: "When I can't sleep, resting in bed is better than nothing."
Coping thought: "Resting in bed is likely to worsen or at least maintain my insomnia. Getting out of bed when I am not sleeping is likely to improve my insomnia."

In cognitive behavioral therapy, patients are asked to develop their own list of automatic thoughts about their sleep habits and come up with reassuring but realistic ways of counteracting them.

STIMULUS CONTROL

Stimulus control is not a technique but a term that refers to situations when a particular behavior (for example, insomnia) is likely to occur in response to a particular stimulus. Certain activities, for example, can produce arousal or wakefulness when people are in the bedroom, whereas others can promote sleep.

Activities that can enhance wakefulness include watching television or reading in bed, or using the computer in the bedroom. Although many people find that these activities help them to relax, most people with insomnia are unaware that the activities could be perpetuating the problem. These activities require attention or alertness and stimulate the brain. The patient begins to associate the activities with insomnia, thus reinforcing the insomnia. In addition, many people become dependent on these activities, fearing that they will not be able to sleep if they do not follow their established routine.

If reading and watching television before bed are an important part of the patient's wind-down period, therapists usually suggest doing them outside the bedroom. This will reduce the likelihood that these behaviors, and the concentration they demand, will become associated with being in bed.

After the lights are out, people should not lie in bed indefinitely trying to sleep; this is another cue that promotes wakefulness. By staying in bed, the person strengthens the association between lying in bed and struggling unsuccessfully to fall asleep, an association that makes it more difficult to restore a normal sleep pattern. Cognitive behavioral therapists encourage patients to get out of bed after they have been lying awake for more than twenty or thirty minutes. Once up, they should engage in a nonstimulating activity until they start to feel drowsy. At that time they should immediately return to bed; this exercise should be repeated as many times as necessary. Sometimes reading a book can work as a relaxant. But if it becomes too difficult to put a good book down before finishing a chapter, patients should consider less engrossing reading materials (for example, a boring history book).

People should not start doing activities that will take a lot of time or be difficult to stop doing, such as paying bills or doing household chores. It can be helpful to keep a notepad beside the bed to jot down any quick thoughts that need to be captured. On a personal note, I find it helpful to get up and write down what I need to do the next day or make a to-do list before going to sleep. All the thoughts keeping me awake are, in a sense, taken out of my head and put on paper. A less stimulating, duller activity will eventually have its effect, but patients should be sure that they do not to return to bed before they feel drowsy.

Therapists will help patients learn about the cues that promote sleep. Establishing a bedtime routine and a regular sleep schedule (or bedtime and wakeup time) are two important cues that help the body transition from wakefulness to sleep. Parents with young children might be used to preparing them for bed with a bath followed by a bedtime story. This is an excellent technique because it sets cues in the child's environment to help signal that sleep is coming. But in the midst of getting children ready for bed and performing the other activities of their extremely busy lives, parents often forget or fail to realize the need to schedule a similar type of bedtime routine for themselves. Doing so is crucial.

In addition to following a routine, people experiencing insomnia need to set a regular bedtime and wake-up time for themselves. Many people fall into the habit of sleeping in on weekends to make up for lost sleep. Although this provides temporary relief, it sets the stage for insomnia.

Most people experience their worst night of sleep on Sunday night. This is partly due to apprehension about returning to work and responsibilities, but in addition, if they have slept longer over the weekend, they find themselves less ready to sleep at the end of it. People who like to stay up a little later on the weekend will not find their sleep overly disturbed, provided they continue to rise at their usual time. Therapists can help patients identify and review the types of cues that could be interfering with their sleep. They might suggest experiments that can help determine whether the patient's sleep improves after removing or adding such cues. Typically, patients need to apply these strategies for several weeks before they start to see a sustained improvement in sleep.

SLEEP RESTRICTION

Another component of cognitive behavioral therapy for insomnia is sleep restriction, a strategy in which the amount of time spent in bed is limited to the time the person is asleep. For example, let's say a person retires at 10:00 P.M. and gets out of bed at 7:00 A.M. but only sleeps from 2:00 A.M. to 7:00 A.M. This person has spent nine hours in bed for five hours of sleep—something sleep specialists would consider terribly inefficient. The patient is asleep for only 56 percent of the time he or she spent in bed.

There are two basic approaches a CBT therapist might suggest to improve this situation: to follow either a strict sleep restriction strategy or a lenient one. In a strict strategy patients calculate their current sleep efficiency and limit their time in bed to actual sleep time. The patient in the example above would go to bed no earlier than 2:00 A.M. for three consecutive nights and continue to get up at 7:00. After three consecutive nights of near perfect (100 percent) sleep efficiency, the patient can begin to go to bed progressively earlier (increasing by thirty-minute increments) until he or she reaches a sleep efficiency of 85 percent or until the sleep is satisfactory.

Many patients find that it is easier to begin this strategy on a weekend, when they have fewer work commitments or other daytime demands. People trying the strategy should expect to feel tired during the day, but after about two weeks the strategy generally has the desired effect.

The lenient approach to sleep restriction involves restricting time in bed more gradually, over a number of weeks, rather than immediately limiting it to match the time spent sleeping. Using the example above, the patient might postpone bedtime until midnight in the first week while continuing to arise at 7:00. In the second week, the person would start going to bed at 1:00 A.M., again, continuing to get up at 7:00. The patient should continue the lenient sleep restriction process until near-perfect sleep efficiency is reached. At that point, he or she would schedule the bedtime progressively earlier, as long as he or she could maintain sleep efficiency. As with the strict strategy, most people will find that the strategy becomes effective fairly quickly, but it will take longer than the strict regimen.

Strategies of using sleep restriction alone have been found to be highly effective even in the primary care setting by a group in New Zealand in 2013.

Relaxation Training

Relaxation training encompasses a variety of relaxation strategies intended to enhance the quality of sleep. When a person is relaxed bodily tension is reduced; additionally, a relaxed person is less likely to focus on his or her sleeplessness. Although self-help books abound with relaxation techniques, here are strategies that have been extensively researched and found to be effective. These relaxation techniques include progressive or deep muscle relaxation, paced breathing, imagery-induced relaxation, and hypnosis or self-hypnosis. Each requires several weeks of daily practice before it can produce results. Patients should practice relaxation and meditation techniques during the day until they have mastered them; only when the techniques have become almost automatic should patients try them at night.

DEEP MUSCLE RELAXATION

Deep muscle relaxation is based on the idea that when their muscles are tensed, people feel unsettled and anxious, and when they are relaxed, people feel calm and peaceful. As developed in *Clinical Behavior Therapy*, by Marvin R. Goldfried and Gerald C. Davison (1976, expanded edition 1994), deep muscle relaxation trains patients to identify when their muscles are tense and to relax them when this happens.

Begin by tensing each of the muscle groups in the list below for about five seconds, and then relax that same muscle group for about ten seconds. When you are tensing, tense the muscles firmly but not hard enough so that you feel pain, cramping, or trembling. You can do this exercise sitting up or lying down, whichever is more comfortable for you. You can do this in bed.

1. Clench your right fist.
2. Clench your left fist.
3. Tighten the biceps muscles in your right arm.
4. Tighten the biceps muscles in your left arm.
5. Bring your right shoulder up toward your ear.
6. Bring your left shoulder up toward your ear.
7. Tighten the muscles of your forehead.
8. Tighten your jaw and grit your teeth.
9. Tighten the muscles in your stomach.
10. Stretch both legs out in front of you, pointing your toes toward the ceiling or sky.

As you release the tension in each muscle group, say the word *relax* slowly to yourself. Focus on the word and on the feeling of relaxation that comes as the tension flows out of your muscles. Tense and relax each group of muscles twice before moving on to the next group.

When you are first learning progressive muscle relaxation, following the instructions on an electronic device might be useful to help you focus. You can record your own instructions by reading into a recording device such as a smartphone. You can even program your smartphone to remind you when you should do the exercise.

PACED BREATHING

Shallow, rapid breathing causes a reduction in blood carbon dioxide levels. It can result in an unsettled, nervous, or even light-headed feeling, caused by mild hyperventilation. The paced breathing technique teaches patients how to breathe slowly and deeply so that they can achieve a deeper state of relaxation. This slow, deep breathing also helps lessen stress, a major culprit in insomnia.

Place one hand on your chest and the other on your abdomen. Inhale slowly through your nose, and as you do so, use the breath to push your abdomen out about one inch. As you exhale, let your abdomen fall back in. Inhale slowly and exhale slowly. Imagine your stomach inflating like a balloon. Allow the hand on your abdomen to rise higher than the hand on your chest (which should move only slightly). Repeat this cycle as many times as needed; make sure you breathe slowly. If you begin to feel dizzy or light-headed, simply breathe through your nose and close your mouth.

After you have finished your breathing exercise, sit quietly for several minutes with your eyes closed. Enjoy the relaxation.

A device that tracks breathing with a smartphone app (2breathe.com) has been approved to treat insomnia.

IMAGERY-INDUCED RELAXATION

Imagery-induced relaxation is based on the idea that imagining a relaxing scene can help people feel more relaxed, breathe more slowly, and feel calmer. Patients are directed to think of a time or place during or in which they felt truly relaxed. Perhaps they were sitting on the edge of a dock at the lake, on the beach, or in their backyard on a hammock, or were out walking on a mild winter day. They might have been at a vacation spot.

Find a comfortable place, close your eyes, and imagine yourself in your relaxing scene. If a particular memory doesn't come to mind, develop a new image that you find relaxing. Try to focus on the smells, sights, and sounds of your image. Let yourself become involved in the image as you get in touch with your senses. Feel free to change any aspect of the image at any time should it cease to be relaxing. After approximately ten to fifteen minutes, open your eyes. Sit quietly and enjoy the relaxation.

HYPNOSIS AND SELF-HYPNOSIS

The principle behind self-hypnosis is that people respond to information that comes from both their conscious and their unconscious minds. Consciously, they can learn to become more relaxed through a number of the exercises already discussed. They can also use their unconscious minds (thoughts and feelings that are outside waking awareness) to learn to identify personal stressors and become more relaxed. Using self-hypnosis, patients can enter a trance-like state of heightened relaxation through any of several techniques.

Therapists who teach self-hypnosis will initially guide the patient through a series of suggestions, including intense focus on some bodily function (such as breathing). Under hypnosis, patients will be guided through various relaxing

scenes and be asked to generate others. The therapist might offer some suggestions about possible ways to approach personal problems. As the patient consid-
ers these suggestions, he or she will continue to hear the therapist talking, but
will pay attention to the therapist only sporadically. The feeling they experience
during hypnosis might be similar to feelings they have had when watching a
very engrossing movie or listening to a favorite song. When the hypnosis session concludes, patients might feel that they have been on a journey of some
kind. After hypnosis, many people spontaneously have new insights about
what has been upsetting them in their lives. Sometimes these insights are not
new; the patient had simply not appreciated their value. The hypnosis experience helps patients relax both by the change in their tension level while in a
trancelike state and by the insights gained regarding what they must do to feel
emotionally better.

Patients should not be concerned that they will lose the ability to control
themselves during hypnosis. They will be in full control and can stop the hypnosis at any time. Typically, the therapist will guide a patient through hypnosis
the first time and then the patient practices it alone at home. It is suggested
that he or she find a comfortable chair or bed for self-hypnosis.

MINDFUL MEDITATION

Mindful meditation (also called mindfulness) is the practice of paying attention in a particular way, in the present moment, nonjudgmentally. The practice of mindful meditation has received a lot of attention in the popular press
and in academic circles in the past decade. Even some police departments
have integrated this technique into their training. Mindful meditation has
been associated with improved sleep in individuals with a variety of health
conditions, and some recent studies have shown an association between regular meditation and lower levels of fatigue among those with insomnia. Medical science does not yet understand precisely how mindfulness helps with
sleep. A number of recent studies have linked mindful meditation to changes
in blood flow to the brain. While this research is very preliminary, it does
support the idea that meditation might lead to changes in the way our brains
process information. Other studies have proposed that mindful meditation
increases our ability to alter thoughts, feelings, and behaviors; clarify values;
deal with situations flexibly; and tolerate unpleasant thoughts and emotions.

Many people also report that practicing mindfulness helps them feel more calm and relaxed.

At its core, mindfulness practice involves non-doing and observing without judgment. It follows a number of principles that, as in relaxation training, are intended to guide the patient's meditation practice but are not hard-and-fast rules. These principles include non-judging, patience, using a beginner's mind, trusting, non-striving, acceptance, and letting go. The principles are described in detail in Jon Kabat-Zinn's *Full Catastrophe Living*. While all these principles are important, many people with health concerns struggle the most with the principles of acceptance and letting go. It is important to realize that acceptance refers to seeing things as they are right at that time, in the present moment. It does not mean accepting that things will always remain this way.

When people think of meditation, many imagine that the goal is to make the mind still or "blank." This is not the goal of mindful meditation. Mindful meditation is a practice in which practitioners focus their attention on their experiences in the present moment in a particular way.

Mindful meditation can be performed while you are eating, brushing your teeth, washing the dishes, or looking after children. When you are practicing with mindfulness, you can let your senses guide your attention. You may notice random or worrisome thoughts pass through your mind, but you don't engage with them. Simply observe them and let them continue passing by. In these moments, you might focus on your breathing. To get the most benefit from these exercises, you should practice them every day for at least two to four weeks. It is very common to feel that nothing is happening at first or to find the meditation boring or frustrating. This is normal; however, most people find that if they persevere and continue to practice regularly these feelings pass. You might wish to try practicing mindful meditation at different times of the day to see what time is best for you. Many people find that starting the day by taking ten to twenty minutes to meditate leaves them feeling calm and refreshed—it's worth getting up early for! Others prefer to include time for meditation as part of their winding-down routine at the end of the day. Some people even use these exercises to calm their bodies and minds when they wake up during the night. There is no wrong way to practice mindful meditation, and the more you can include it in your day, the more quickly you will notice changes.

Tapering Off of Sleeping Pills

As we have seen, one of the main advantages of cognitive behavioral therapy is that it enables the patient to stop using sleeping pills. Many people prefer to sleep without medication, but some feel they cannot do so. Whether a patient takes medication or not is a matter of personal preference. But for those who worry that they are developing a psychological dependence on sleeping medications (and sometimes a physical dependence as well), CBT can offer an alternative therapy for insomnia. Some patients wean themselves gradually: one woman going through behavioral treatment for a sleep problem said that she liked to have her sleeping medication on hand just in case she needed it. She felt apprehensive and anxious unless it was available. But as noted above, taking a pill for sleep reinforces the idea that the patient will not be able to sleep without it. (Sleep medication can also be expensive.) Therapists work with patients to help them taper off the sleeping medication once they have learned other skills to manage their insomnia. If a patient has been taking a sleeping pill regularly for several years, it could be dangerous to stop cold turkey. This is why the patient needs to taper off the medication.

To taper off a sleep medication, start by reducing your medication to its lowest dose and taking it as often as you did the full dose. If you have been on sleeping pills for many years, it is safest to do the tapering under the supervision of a doctor; the tapering process might last several months. After you have adjusted to taking only a small dose, you should begin scheduling times when you will not take a sleeping pill, no matter how hard it is to sleep. It is best to set aside at least two nights for this. Many people choose to begin on a weekend or at a time when they have fewer responsibilities to deal with during the day. Most find that they feel anxious and have difficulty sleeping the first night without sleeping pills, but by the second night they are usually tired enough to sleep well unassisted. It typically takes several weeks to stop using sleeping pills altogether, and people often have lapses or difficult times during this period. Many people report feeling a sense of intense pride and accomplishment once they have managed to get off their sleeping medication.

A study published in 2003 reported that the combination of cognitive behavioral therapy and tapering off medications was much more effective than tapering off alone in helping people stop using sleeping pills. After one year, 70 percent of those who received the combined treatment had stopped using sleeping pills, compared to 24 percent of those who tapered off without CBT.

Back to the Woman Who Was Afraid to Make a Fool of Herself

Since the patient's insomnia stemmed from her embarrassment about her sleep-walking and screaming, I referred her to a psychologist for cognitive behavioral therapy. Several months later, I received a letter from the psychologist reporting that the patient had significantly improved. She was no longer afraid of what would happen after she fell asleep, and this had contributed to her getting a better night's sleep with fewer episodes of sleepwalking and sleep terrors.

Many patients with insomnia want to avoid medications, fearing that these could become addictive or dangerous. They begin to think of their insomnia as an untreatable condition. But therapies and therapists are available to give people a good night's sleep without medication.

20

Medications That Treat Sleep Disorders

THE MYSTERY. Many medications affect sleep. Some have a side effect of insomnia; some help people with insomnia get to sleep. These medications can have unwanted effects.

~~~~~~~~~~~~~~~~~~~~~~

The Case of the Truck Driver with Insomnia

Doctors always remember the cases that were easily solved simply because they asked the right questions. One such case involved a woman in her mid-thirties who had had severe insomnia for several years before she came to see me. She was concerned because she drove a truck for a living, delivering soft drinks for a major bottling company, and she was having trouble staying alert in the daytime. She told me that it took her several hours to fall asleep and that she frequently woke up at night and had trouble falling asleep again. Though she estimated that she probably slept only two to four hours each night, she did not want to take sleeping pills or any other medications. She was at her wit's end.

We went through her medical history, but I could not find anything obviously wrong. She neither smoked nor drank alcohol or coffee. She had never had any medical problems and did not use any medications. I was stumped. However, the solution to her problem became clear when I asked her how the sleepiness was affecting her job. She told me that she was very sleepy during the day, but she perked up when she drank the soda she delivered—and she drank about ten quarts a day.

Drugs People Take to Fall Asleep

Many people depend on medications to help them fall asleep. Nothing else has worked for them, and they feel that their lives and health are better when they are on medication than when they are off it. There are hundreds of products being marketed to help people sleep. Several are effective and safe, and medications with few side effects have been introduced.

The products available to help people sleep can be categorized into four types.

Prescription drugs (hypnotics). Hypnotics are drugs that have been scientifically tested and released for use specifically to treat insomnia, and have been approved in the United States by government regulatory agencies.

Prescription medications used off-label. Medications that have been scientifically tested and approved by the Food and Drug Administration for treating a specific disease (not insomnia), but have sleepiness as a side effect, are sometimes prescribed as medication to help people sleep. The most common medications prescribed for this purpose are the antidepressants.

Over-the-counter products. Hundreds of nonprescription sleeping pills are available in retail outlets such as drug stores, and some can be effective in some cases.

"Natural" products. Drug or health food stores sell a variety of herbal and other naturally occurring products that are reputed to promote sleep.

PRESCRIPTION DRUGS (HYPNOTICS)

Most people who try a sleeping pill think that all they have to do is swallow it and like magic it will put them to sleep. If only it were so simple. What actually happens is that first the pill has to dissolve in the intestinal tract, next be absorbed into the body through the bloodstream, then pass through the liver, where it might be broken down, and once more make its way into the bloodstream until it finally reaches the brain, where it attaches to receptors. And throughout this process, the body will try to rid itself of the chemical. Before taking a sleeping pill, people should be aware of the basic principles of how it works.

Different drugs have different onset periods: the length of time it takes for the drug to go from the stomach into the body and finally into the brain varies by the drug and by the individual.

A medication will not put the user to sleep until molecules of the drug have activated a certain number of sleep receptors in the brain.

The drug will continue to keep the user asleep as long as the required number of sleep receptors continue to be activated.

Drugs vary in the way they are broken down by the body and therefore have different durations of effect. Scientists measure how long the effects of a drug last by finding out how long it takes for half of the chemical to disappear from the body (its half-life).

Thus, if a drug has a half-life of two hours, its level in the blood will have dropped to one-half in two hours, to one-fourth in four hours, and to one-eighth in six hours. However, some of the drug can still be attached to receptors in the brain, or the drug might have changed the receptor in such a way that the effect continues long after the drug is gone from the body. Another complication is that when some drugs are broken down, byproducts can be present that have the same effect as the drug.

The types of chemicals used in prescription sleeping pills that affect sleep receptors changed in the 1970s. Older drugs, including medications called barbiturates, were used to affect other receptors in the brain, and sometimes an overdose could cause death. These drugs are rarely used today as sleeping pills, although I occasionally encounter a person who still uses them. Barbiturates were then replaced by benzodiazepines, a large number of which are still on the market (for example, triazolam, flurazepam, and temazepam). These medications attach more specifically to sleep receptors, but they have other

FDA-Approved Medications to Treat Insomnia

Nonbenzodiazepines

Medication	Brand name	Time to Peak (in hours)	Half-Life (in hours)
Zaleplon	Sonata	1	1
Zolpidem	Ambien	1.6	2.5
	Intermezzo	0.6–1.25	2.5
Eszopiclone	Lunesta	1	5–7
Doxepin	Silenor	3.5	17
Ramelteon	Rozerem	0.75	1–2.6
Suvorexant	Belsomra	2	12

Benzodiazepines

Medication	Brand name	Time to Peak (in hours)	Half-Life (in hours)
Triazolam	Halcion	1–2	1.5–5.5
Temazepam	Restoril	1.2–1.6	3.5–18.4
Flurazepam	Dalmane	0.5–1	47–100
Quazepam	Doral	2	47–100

Note: Some of the branded products in the table are available in generic forms. The data in the table were compiled from several sources, including the U.S. National Library of Medicine, the National Institutes of Health, articles in medical journals, and product inserts. Some companies may no longer be marketing some of the products in some countries.

effects as well. Lethal overdoses of benzodiazepine medications are rare, but they can occur in combination with other drugs or alcohol. The most recently introduced hypnotics, called nonbenzodiazepines, target more specifically the sleep receptors with little effect on the rest of the brain. Overdoses of these drugs are rarely lethal.

The table above lists the more commonly used and more recently approved medications that act on the sleep receptors. (All prescription medications have two names, a chemical name and a brand name.) The longer the half-life of a medication (the longer the drug takes to be eliminated from the body), the more likely the user is to have residual sleepiness or drowsiness the following morning, which can impair his or her ability to perform certain activities such as driving. (Intermezzo, taken under the tongue, starts working quickly and is effective for middle-of-the-night insomnia as long as it is taken when more than four hours of bedtime remain.) Notice that two of the drugs have a half-life of more than twenty-four hours. After repeated use, such drugs will probably accumulate in the body and cause daytime sleepiness or grogginess. Patients may not even realize that they are impaired if they have been using such medications for long periods of time.

Some of the drugs can cause disturbing, bizarre behaviors during sleep such as sleep eating and sleepwalking; some users have even gone outside or driven a motor vehicle. These occurrences should be reported immediately to the doctor.

Research suggests that among the benzodiazepines, the shorter the half-life, the more likely it is that there will be side effects (such as temporary worsening of insomnia) when the person stops using the medication. Generally, both the benzodiazepines and nonbenzodiazepines are considered fairly safe because they do not depress breathing if taken in the normal doses (those suggested on the label). Patients with a breathing problem or a serious medical problem should make sure the doctor is aware of it before using any sleeping medication.

PRESCRIPTION MEDICATIONS USED OFF-LABEL
Some readers of this book may have noticed that their sleeping medication is not listed in the table. This is because the doctor might have chosen to treat their insomnia with medications that are normally used for other conditions but have sleepiness as a side effect. In the United States, such medications have not been approved as treatments for insomnia by the FDA, and this practice is called off-label prescribing. This practice is perfectly legal, but insurance carriers may not cover these medicines. Some antidepressants, for example, have sleepiness as a side effect and have frequently been used to treat sleep problems in certain patients, especially those who also have symptoms of depression. This dual efficacy may help explain why so many women, who are considered more prone to depression, are on antidepressants for insomnia. The most widely used antidepressants are trazodone (Desyrel) and some of

the older medications such as amitriptyline, imipramine, and doxepin. A very low dose of doxepin has been approved as a hypnotic by the FDA under the brand name Silenor.

OVER-THE-COUNTER PRODUCTS

Most of the sleep aids or medications found on drugstore or grocery store shelves to treat insomnia have an antihistamine as their main ingredient. Many of these products were originally introduced to treat allergy problems, and their main side effect was sedation. The newer antihistamines do not have this side effect, so people who buy one of the new antihistamines to help them sleep will be disappointed unless their problem sleeping is that allergies are giving them a stuffy nose and an itchy throat. For these sufferers, getting relief from these symptoms will help them sleep.

The over-the-counter sleep products that contain the older antihistamines have not been extensively or rigorously tested for their effect as sleep aids, but they are probably safe for short-term or occasional use. Their main side effect is also their main effect: sedation. People often feel dopey or groggy the next day. This is because the medication might not have completely cleared from the body, or it might not have given the best type of sleep. Use of these medications for weeks, months, or longer is not advised because they affect the histamine system in the body, which can cause unwanted effects, including restless legs syndrome, nervousness, nausea, and more. People using these drugs should read the fine print on the package. As a general rule, I do not recommend sleep remedies that are manufactured for other therapeutic purposes. My caution is based on the fact that most of these drugs have not been adequately tested for any but their stated purpose.

Some products that promise to help sleep might contain a mixture of compounds including antihistamines, melatonin, and other ingredients. These products have not been proven to be effective.

Some people self-medicate with over-the-counter sleeping pills so they can sleep through their bed partner's snoring. For better ways to deal with a snoring bed partner, see Chapter 12.

"NATURAL" PRODUCTS

Several products that are sometimes used to treat sleeplessness are available in health food and other retail outlets. These include cannabis, melatonin, kava, and valerian root. Some also swear by chamomile tea. These products were not tested as rigorously as the hypnotic medications that have been approved by government agencies and are prescribed by a doctor. The long-term effect of these products and the ways they interact with other medications are generally not known. People considering these natural products should be aware that much less scientific information is available about these products than about prescription sleep medications. As I noted in Chapter 5, the fact that a product is "natural" does not mean it is safe, nor does it mean that the product is what it is labeled to be. Using DNA testing, a recent study found that products from only two of twelve manufacturers contained the ingredients that were listed on the labels. Products of the other ten companies had ingredient substitutions, contamination with other plants, or fillers that could pose serious health risks. Buyer beware!

Even when we know that the ingredients are pure and as listed, we do not know whether taking such products over extended periods is safe or effective. Many people complain that natural remedies do not work for them, although many others who take these products find them satisfactory. The most important issue I want to emphasize here and throughout this book is that anyone who has a serious problem falling or staying asleep that lasts more than a few weeks needs to see a medical practitioner to make sure that the insomnia is not a symptom of another disorder. Sleeping pills do not cure a single medical condition. People should be very careful about using any medication to treat insomnia, particularly if they find themselves using it every night for more than a few months.

Cannabis. For people in severe pain that is interfering with sleep, cannabis has been reported to reduce pain and improve sleep. In some jurisdictions doctors may prescribe marijuana for medical use. When patients stop using the marijuana, however, it can result in disturbed sleep. In a few U.S. states recreational use has been decriminalized. There is more than one species of marijuana plant (sativa, indica), and different species have different concentrations of ingredients; additionally, plants have been genetically modified to vary the amounts of these chemicals. Users appear to favor some preparations more than others to treat pain or insomnia. In the absence of another reason to prescribe marijuana (pain), I do not recommend its use as a sleep aid.

Melatonin. Melatonin is a hormone produced by the human brain in the pineal gland. It has been called the hormone of darkness because sunlight brings a drop in the level of this hormone. Most people who take melatonin to fall asleep have either insomnia or a problem with their body clock (see Chapter 8), usually as a result of jet lag or crossing time zones. In many countries it is widely available in health food stores, drugstores, and other retail outlets. In some countries a pharmaceutical formulation is available that requires a doctor's prescription.

Melatonin is one of only two hormones available without a prescription in the United States. (The other is DHEA, a product with male hormone properties that is frequently used by athletes to bulk up their muscles.) Although people argue that melatonin is natural because it is a chemical that the brain naturally produces, the dosage usually taken is many times greater than that produced by even the most high-functioning pineal gland.

Whereas most prescription drugs undergo rigorous testing to make sure they are effective and safe, melatonin has not been studied to find out whether it is safe when used over the long term by the general population. Additionally, no rigorous studies have been performed to determine whether melatonin is effective in the treatment of insomnia or what the optimal dose should be. We do not know what an effective dose would be for most people. Furthermore, people buying it over the counter should be aware that the packaging usually does not list information about side effects—and they might actually feel dopey and tired the day after using it.

Though there have been some studies as to melatonin's effectiveness in alleviating both jet lag and delayed sleep phase, the studies have not involved a large number of subjects. The fact that it is a natural substance does not mean that it is safe to use. Both insulin and thyroid hormone are natural substances, but they will never be approved in the United States for widespread use without a prescription because we know that taking too much of these hormones can cause severe medical problems. Also, unlike the companies that produce prescription drugs, manufacturers and importers of melatonin are not closely regulated by the FDA to ensure that the manufacturing process is safe and that the ingredients are accurately stated on the label or product insert. Melatonin is not classified as a drug in the United States; it is marketed as a dietary supplement along with vitamins and similar products. Melatonin is available because of a technicality in the U.S. drug laws, and I believe that it needs to undergo the same rigorous study that we would give to any other drug.

Two drugs, ramelteon (Rozerem) and tasimelteon (Hetlioz), that stimulate some of the same receptors as melatonin have been approved by the FDA. These medications are very safe, but they are not as effective as other approved prescription hypnotics. Tasimelteon has been approved in the treatment of people with circadian rhythm problems, in particular people who are blind and as a result cannot synchronize their circadian clock.

Kava and valerian. Two other products available over the counter, especially in health food stores, are derived from the plants kava (a plant that grows in the South Seas) and valerian (a flowering plant).

Few medical studies have been published about the effectiveness of kava for treating insomnia. What is known, however, is that kava can, in rare cases, cause liver failure. This product has been removed from the market in Canada because it was not shown to be effective and could pose a dangerous health risk. In the United States, the FDA has issued an alert warning the public about this serious side effect.

Valerian has been studied a bit more, but again the sampling of people studied using modern methods has in most cases been quite small (fifteen to thirty people), and the results are inconsistent. Some studies showed sleep-promoting benefits, while others showed little or modest improvement. The largest scientific review of all the articles about valerian (and other herbal products) found that the results were inconclusive.

Medicine has learned the hard way that side effects of drugs, even when rare, can be dangerous. These risks should not be ignored or dismissed. How can a conscientious doctor recommend a treatment when a study has examined only a small number of subjects?

Drugs People Use to Stay Awake

Feeling sleepy during the daytime is a common consequence of our lifestyle, the distractions in the world we live in, and a number of sleep disorders. Many sleep-deprived people cannot function without their morning coffee, and at the office, the coffeemaker has become as ubiquitous as the computer. Caffeine and other stimulants increase the function of several organ systems, including the brain and the cardiovascular system. At low doses only one organ—for example, the brain—might be stimulated and cause the user to be more alert. At higher doses several systems might be stimulated, resulting in unwanted side effects.

CAFFEINE

Caffeine is probably the most commonly used stimulant drug in the world. It has been estimated that at least 80 percent of North Americans take some form of caffeine during the course of a day. Caffeine is found in soft drinks, some foods, and of course coffee and tea, as well as some medications. (The website Caffeineinformer.com gives the caffeine content of a number of products.) Many people drink a cup of coffee or tea to become more alert first thing in the morning, and they might have two or three more cups during the day. But although caffeine can help them start the day, it carries risks.

After someone drinks coffee (it doesn't really matter how much), it takes the liver three to four hours to reduce the caffeine blood level by half. Thus it could take nine to twelve hours for the caffeine to completely clear the system. Birth control pills can slow down the body's elimination of caffeine even further.

People who drink more than 200 milligrams of caffeine a day, especially in the afternoon or evening, are likely to have insomnia. As we saw in Chapter 1, while we are awake we accumulate a chemical called adenosine in the brain that makes us sleepy. Caffeine antagonizes the effect of adenosine. Few people should take more than 400 milligrams of caffeine a day. The amount of caffeine in dark roasts varies, but two cups of most of them have more than 400 milligrams. Even many "decaf" coffees have some caffeine, as reported in *Consumer Reports* in 2014. (As regards coffee, the news is not all bad: coffee—even decaf—has been reported to have many beneficial effects, including protecting the liver from scarring; it might also have anti-cancer properties and help counter Parkinson's disease and dementia.)

"Energy drinks" and related products might also contain a great deal of caffeine. I have had patients who drank energy drinks or other products that promised hours of energy all day long and then developed symptoms of caffeine toxicity. Such products can even result in death if overused.

Menopausal women should be aware that medical studies suggest that more than 300 milligrams of caffeine a day (eighteen ounces of brewed coffee) speeds up bone loss and can increase the risk of osteoporosis.

In excess amounts, caffeine also contributes to symptoms of anxiety.

Parents might be amazed (and dismayed) to discover how much caffeine their children imbibe without their realizing it, particularly in soft drinks. Some of the most popular soft drinks contain high levels of caffeine. A study published in 2003 reported that caffeine intake was significant in seventh-, eighth-, and ninth-graders; 70 percent of them drank caffeine daily, and almost 20 percent were drinking more than 100 milligrams a day. One eighth-grader took 380 milligrams a day! The 2004 National Sleep Foundation Poll reported that 26 percent of children over three years old have at least one caffeinated drink daily. Children drinking caffeinated beverages sleep less than those who do not, averaging 9.1 versus 9.7 hours per night, and thus losing about 3.5 hours a week.

During pregnancy the body breaks caffeine down more slowly. Pregnant women have many reasons to feel sleepy (see Chapter 4) and might believe that they could combat the sleepiness by drinking coffee. Research from Sweden published in 2000 showed that women who consumed 300 to 499 milligrams of caffeine per day increased their risk of miscarriage by 40 percent, while those who consumed 500 milligrams or more per day increased their risk by 120 percent. Another study, from Denmark, in 2003 reported a 120-percent risk of miscarriage when caffeine use exceeded 375 milligrams per day. How much coffee is safe in pregnancy? Probably about one cup per day.

Over the years I have seen many patients, including children, with insomnia caused by excessive caffeine. Some people even become addicted to caffeine, drinking fifteen to thirty cups a day. Cutting back is difficult for them and they might develop symptoms such as headaches and nervousness when trying to reduce their intake. The best way for caffeine addicts to reduce their dependence is slowly, over one or two weeks.

PRESCRIBED STIMULANT MEDICATIONS

Some stimulant medications are prescribed to treat sleepiness in patients with sleep disorders such as narcolepsy (see Chapter 13); they are also used to treat attention deficit hyperactivity disorder in adults and children. These medications have been used for decades. Stimulant medications work by affecting cells in the central nervous system. Some stimulant medications—which include amphetamine, dextroamphetamine, methamphetamine, cocaine, and methylphenidate—have also become street drugs. They affect the sympathetic nervous system, so they can cause changes in heart rate and rhythm and might increase blood pressure and cause jitteriness. They can increase the levels of the chemical dopamine in the brain and elsewhere, which can excite nerve cells, including those controlling heart rate and blood pressure.

Methylphenidate and related drugs (Ritalin, Concerta, Vyvanse) are widely used to treat children with ADHD. They are also sometimes used to treat narcolepsy. But though these can be effective in combating sleepiness in narcolepsy, they seem to have the opposite effect on ADHD. Other drugs used to treat ADHD include guanfacine (Estulic, Tenex, and the extended-release Intuniv), which is also used to treat high blood pressure, and atomoxetine (Strattera).

The amphetamine drugs include amphetamine, dextroamphetamine (Dexedrine, DextroStat), and methamphetamine (Desoxyn). Adderall and Biphetamine are combination drugs that contain both amphetamine and dextroamphetamine. Years ago, doctors prescribed amphetamines as appetite suppressants; many women, in particular, took them to lose weight. One doctor told me that her mother had taken amphetamine around 1950 to lose weight while she was pregnant with her. The long-acting versions of these drugs are sometimes used to treat ADHD. Stimulants are more difficult to prescribe than other types of drugs because of the rigid controls on medications that have abuse potential. Pemoline (Cylert) is a stimulant medication that is no longer available in the United States or Canada because of concerns about side effects, especially liver problems. Mazindol (Mazanor and Sanorex), which is used to decrease appetite, is also sometimes prescribed as a stimulant.

Many students who do not have sleep disorders are somehow obtaining prescription stimulant medications and using them to try to improve their school grades.

Ironically, the main effect of and reason why a stimulant is used, alertness, is also its major negative side effect. The user's alertness might continue long into the night, making it very difficult for him or her to fall and stay asleep.

ALERTNESS-PROMOTING DRUGS

Modafinil (Alertec in Canada, Provigil in the United States and the United Kingdom) was introduced in North America to treat sleepiness in patients suffering from narcolepsy. A modified longer-acting version of this compound called armodafinil (Nuvigil) has also been approved. Modafinil works differently from the stimulants mentioned earlier because it does not stimulate several organ systems but instead seems to act on the centers of the brain that are involved in keeping the person awake. It is beginning to be used for many conditions in which the patient might be sleepy or tired, including multiple sclerosis, depression, Parkinson's disease, and cancer. It has been approved in the United States for use in treating people with sleepiness caused by shift-work disorders and in treating sleep apnea patients who are using CPAP but have residual sleepiness. This is a medication that seems to wake people up without the sympathetic nervous system activation that is observed in those who take amphetamines. Provigil is often best taken twice a day, the first dose first thing in the morning, and the second dose at lunchtime. If patients take the medication later in the day, they might have trouble sleeping that night. Nuvigil is taken once a day, in the morning. These new compounds so far have not been shown to be addictive and have few side effects.

Sodium oxybate (a form of gamma hydroxybutyrate), by enhancing sleep for narcolepsy patients, improves alertness in these patients. It is not approved to be used for any other condition at this time.

Back to the Truck Driver with Insomnia

The truck driver was drinking about ten quarts a day of a cola containing caffeine. Her insomnia was caused by the huge amounts of caffeine she was unknowingly taking into her body. Although she got the soft drinks free, she had paid a big price over several years. She had never realized that the soft drinks had so much caffeine. After she weaned herself from her dependence on the cola and limited her intake of caffeine to a reasonable amount, her problem was solved.

Medications used to treat sleep problems can improve symptoms, but they can have unwanted effects. Medications can be used safely only when the doctor and the patient know what the problem is and have discussed treatment options. For insomnia patients who find CBT ineffective or difficult, medication taken in consultation with a doctor can be the most acceptable option.

21

Time for Bed

THE MYSTERY. Restful sleep is something that everyone needs and seeks. It is essential to health and and helps us lead happy and productive lives. A good night's sleep is as important as the air we breathe and the water we drink.

~~~~~~~~~~~~~~~~

If you are reading this book in bed, I hope you'll be able to close it at the end of this chapter and feel reassured that the solution to your sleep disorder will soon be within your grasp.

Sleep disorders and lack of sleep seriously affect many people's lives. When someone cannot sleep properly at night, he or she feels sleepy throughout the day, has no energy to get important tasks done, and is at risk for a number of health problems. Sleep problems can affect us all, but they do not affect us all in the same way. Women, men, and children have different sleep issues.

Females differ from males in many aspects of life, and sleep is no exception. Women's sleep can be negatively affected by hormones, pregnancy, menopause, and juggling outside work with the job of being primary caregiver at home (men who are the primary caregivers face the same challenges). For women of childbearing age, each month brings hormonal changes that can cause sleep disturbances. A common problem in hormone secretion, polycystic ovarian syndrome, can lead to a dangerous problem, sleep apnea. And when a woman becomes pregnant, a dramatic change occurs in her physiology and anatomy to prepare for the miracle of birth. The mother pays a price for this miracle. About 80 percent of pregnant women have disturbed sleep, and some develop sleep apnea and movement disorders. Making sure that the mother-to-be does not become deficient in folic acid and iron might improve her sleep and protect the developing baby from a permanent neurological condition.

Another big change in hormone levels occurs during menopause, which affects a woman in many ways. The reduction in the normal levels in female hormones increases a woman's risk of developing sleep-breathing problems and heart disease. It also causes severe menopausal symptoms, such as hot flashes, which can seriously disrupt sleep. More than half of menopausal and postmenopausal women have insomnia. Though there are treatments, the best way to treat sleep problems during menopause remains unclear. If you are a woman with one or more of the sleep problems described in this book, you should be able to find here the information you need to seek medical help specifically directed to your needs.

For men, sleep problems can come from stresses related to work and their lifestyle. Because they tend to snore more than women, they are considered at special risk for sleep apnea. (It can also make life miserable for their bed partner, putting a strain on the relationship.) As they age they can develop sleep problems related to heart disease and stroke.

Children are not merely small adults. They need more sleep than adults. They can also get sleep apnea related to enlarged tonsils, a small jaw, or obesity. As they age they are prone to develop problems with their body clocks, and the mid-teenage years is the period when narcolepsy usually manifests. Sleep problems in children can lead to academic failure, which can affect or even ruin their entire life.

One of the most potentially dangerous—and common—sleep disorders is sleep apnea. We now know that sleep apnea affects about 4 percent of all adults. What is less well known is that in many cases a cure can be simple, as was illustrated by the case of the farmer's fourteen-year-old daughter whose life was almost ruined by her big tonsils. Women's sleep apnea symptoms sometimes differ from those of men; consequently, women suffering from sleep apnea are often misdiagnosed and treated for depression. As we saw, sleep apnea is considered by many doctors, as well as the general population, to be a disorder of overweight men; however, many women and children have sleep apnea, and they might not be obese. The information in this book can help you recognize when you or a family member should seek professional help and suggest treatments that have been proven to work.

Sleep apnea is often suspected because the patient snores. As we have seen, snoring can not only alert patients to a potential danger to themselves; it can cause serious secondhand sleep problems for bed partners. The strategies I have suggested for dealing with snoring can help both bed partners have healthier and more restful sleep.

The amount of sleep your body needs is as individual as your hair color. Different amounts of sleep are normal for different people. Additionally, the "clock" in the brain that controls sleepiness and alertness can run on a different time from the clocks of those around you. When your clock is different from that of the general population, you may either have trouble staying asleep all night or be unable to fall asleep. Travel across time zones can seriously confuse this clock. Information in this book can alert you to what can go wrong with your body clock and how you can reset it.

The most common sleep problem for both men and women is insomnia, though it is much more common among women than among men. We have seen that insomnia is not just a disease; it is a symptom of something else that has gone wrong in the body—a medical, psychological, or psychiatric condition, or a reaction to stress. Medical diseases ranging from heart failure to diabetes, acid reflux, ulcer disease, arthritis, and many other painful conditions, including cancer, can lead to severe insomnia. We have seen how these conditions can cause insomnia and what means are available to help with insomnia. If you have persistent insomnia, you need to consult your doctor so that he or she can recommend a specific treatment for the medical or psychiatric condition causing it.

Restless legs syndrome and other movement disorders affect up to 15 percent of the adult population worldwide. In many cases, the underlying cause of RLS is iron deficiency. Women are much more likely to become iron deficient in their lifetime than men because of their menstrual cycles, and RLS frequently starts during pregnancy. Information in this book can help you recognize this condition in yourself and family members. Fortunately, we now know that certain simple treatments can often cure this miserable disorder.

I have written repeatedly in this book about the great power the brain and nervous system wield over the quality and quantity of our sleep. Nearly all psychiatric disorders have disturbed sleep as a characteristic feature. The most common of these disorders is depression, which affects about 5 to 10 percent of adults worldwide. Women are more likely than men to be treated for depression, are therefore more likely to have sleep problems related to depression, and are more likely to be on antidepressant medications, many of which can cause disturbed sleep. Along with antidepressants, the effect of drugs on sleep can be either good or bad. Many sleep remedies, such as prescription drugs, over-the-counter medications, and products obtained in health food stores, have potentially negative effects on sleep. Many medications, both prescribed and not prescribed, can worsen an already existing sleep problem and can even cause sleep problems. Because North American women statistically take almost twice as many drugs as men, they are more likely to suffer from these unwanted effects.

Other brain-controlled sleep problems can disturb a person's sleep and result in a fear of going to sleep. This can occur in combat veterans and others suffering from posttraumatic stress disorder. These sleep problems include sleepwalking, sleep talking, sleep paralysis, hallucinations, and even violent physical behavior. In one disorder, the sleeper reacts to dream content and might injure the bed partner. Though this disorder is much more common among males, it is the bed partner—often a woman—who can suffer the physical consequences. This problem too can usually be treated quite easily. The information in this book should help you distinguish between sleep behaviors that are risky (to yourself or others) and those that are relatively harmless.

Even serious conditions like narcolepsy can often be treated, allowing the sufferers to live normal lives. The hardest thing for many narcolepsy patients is to receive a correct diagnosis. The symptoms described in this book should help you recognize this devastating disorder in yourself or a family member and get the help you need.

From the extreme of narcolepsy to something as simple as drinking too much coffee, I've tried to catalogue the many problems that can disrupt your sleep. If after reading this book you realize you have a sleep problem, the good news is that there is help available. There are clinics with doctors devoted to people with sleep problems all over the world.

Armed with the knowledge gleaned from this book and a precise description of your symptoms, you will be able to help your doctor help you.

There are still many mysteries about sleep, but we now understand a lot of them. I wrote this book to give you the information and the tools you need to awaken every morning feeling rested, wide awake, and ready to tackle the world.

I wish you sweet dreams.

BIBLIOGRAPHY

〜〜〜〜〜〜〜〜〜〜

CHAPTER 1: WHY DO WE SLEEP?

Aserinsky, E., and N. Kleitman. "Regularly Occurring Periods of Eye Motility, and Concomitant Phenomena, During Sleep." *Science* 118, no. 3062 (1953): 273–74.

Avidan, Alon Y. "Normal Sleep in Humans." In *Atlas of Clinical Sleep Medicine*, 2nd ed., ed. Meir H. Kryger, Alon Y. Avidan, and Richard B. Berry, 70–97. Philadelphia: Elsevier/Saunders, 2014.

Ayas, N. T., et al. "A Prospective Study of Sleep Duration and Coronary Heart Disease in Women." *Archives of Internal Medicine* 163, no. 2 (2003): 205–9.

"Bill Clinton Sleeps Through King Tribute." *Telegraph* (London), January 22, 2008, available at http://www.telegraph.co.uk/news/worldnews/north america/usa/1576275/Bill-Clinton-sleeps-through-King-tribute.html.

Ekirch, A. Roger. *At Day's Close: Night in Times Past*. New York: Norton, 2005.

Hirshkowitz, M., et al. "National Sleep Foundation's Updated Sleep Duration Recommendations: Final Report." *Sleep Health: Journal of the National Sleep Foundation* 1, no. 4 (2015): 233–43. doi:10.1016/j.sleh.2015.10.004.

Huffington, Arianna. *The Sleep Revolution: Transforming Your Life, One Night at a Time*. New York: Harmony, 2016.

Konnikova, Maria. "Goodnight. Sleep Clean." *New York Times*, January 11, 2014. Available at http://nyti.ms/1c5pkqN.

"Normal Sleep and Its Variants." In Meir H. Kryger, Russell Rosenberg, Douglas Kirsh, and Lawrence Martin, *Kryger's Sleep Medicine Review: A Problem-Oriented Approach*, 2nd ed., 1–20. Philadelphia: Elsevier, 2015.

Siegel, Jerome. "Sleep in Animals and the Phylogeny of Sleep." In *Atlas of Clinical Sleep Medicine*, 2nd ed., ed. Meir H. Kryger, Alon Y. Avidan, and Richard B. Berry, 65–69. Philadelphia: Elsevier/Saunders, 2014.

Yetish, G., et al. "Natural Sleep and Its Seasonal Variations in Three Pre-Industrial Societies." *Current Biology* 25, no. 21 (2015): 2862–68. doi:10.1016/j.cub.2015.09.046.

CHAPTER 2: SLEEP REQUIREMENTS IN THE LIFE STAGES

"2006 Sleep in America Poll, Summary of Findings: Teens and Sleep." National
    Sleep Foundation. http://sleepfoundation.org/sites/default/files/2006_sum
    mary_of_findings.pdf (accessed August 25, 2016).
Bloom, H. G., et al. "Evidence-Based Recommendations for the Assessment and
    Management of Sleep Disorders in Older Persons." *Journal of the American
    Geriatric Society* 57, no. 5 (2009): 761–89.
Brand, S., et al. "Associations Between Infants' Crying, Sleep and Cortisol Secre-
    tion and Mother's Sleep and Well-Being." *Neuropsychobiology* 69, no. 1 (2014):
    39–51. doi:10.1159/000356968.
Centers for Disease Control and Prevention. "Sleep Duration and Injury-Related
    Risk Behaviors Among High School Students." *Morbidity and Mortality
    Weekly Report* 65, no. 13 (April 8, 2016): 337–41. http://www.cdc.gov/mmwr/
    volumes/65/wr/mm6513a1.htm.
"Five Clusters of Sleep Patterns." National Sleep Foundation. https://sleepfoun
    dation.org/sleep-news/five-clusters-sleep-patterns (accessed August 25, 2016).
Hirshkowitz, M., et al. "National Sleep Foundation's Updated Sleep Duration
    Recommendations: Final Report." *Sleep Health: Journal of the National Sleep
    Foundation* 1, no. 4 (2015): 233–43. doi:10.1016/j.sleh.2015.10.004.
Mindell, J. A., et al. "Bedtime Routines for Young Children: A Dose-Dependent
    Association with Sleep Outcomes." *Sleep* 38, no. 5 (2015): 717–22. doi:10.5665/
    sleep.4662.
Moon, R. Y., et al. "Safe Infant Sleep Interventions: What Is the Evidence for
    Successful Behavior Change?" *Current Pediatric Reviews* 12, no. 1 (2016):
    67–75.
Rosenberg, R., et al. "2011 Sleep in America Poll: Communications Technology
    in the Bedroom." National Sleep Foundation, March 7, 2011. https://sleep
    foundation.org/media-center/press-release/annual-sleep-america-poll-explor
    ing-connections-communications-technology-use- (accessed May 1, 2016).
Yang, L., et al. "Longer Sleep Duration and Midday Napping Are Associated
    with a Higher Risk of CHD Incidence in Middle-Aged and Older Chi-
    nese: The Dongfeng-Tongji Cohort Study." *Sleep* 39, no. 3 (2016): 645–52.
    doi:10.5665/sleep.5544.

CHAPTER 3: THE REPRODUCTIVE YEARS

Baker, Fiona C., and Louise M. O'Brien. "Sex Differences and Menstrual-
    Related Changes in Sleep and Circadian Rhythms." In *Principles and Prac-
    tice of Sleep Medicine*, 6th ed., ed. Meir H. Kryger, Thomas Roth, and Wil-
    liam C. Dement, 1516–24. Philadelphia: Elsevier, 2017.

Baker, Fiona C., et al. "Sleep Quality and the Sleep Electroencephalogram in Women with Severe Premenstrual Syndrome." *Sleep* 30, no. 10 (2007): 1283–91.

Ehrmann, D. A. "Metabolic Dysfunction in PCOS: Relationship to Obstructive Sleep Apnea." *Steroids* 77, no. 4 (2012): 290–94. doi:10.1016/j.steroids.2011 .12.001.

Lee, Kathryn A. "The Menstrual Cycle." In *Atlas of Clinical Sleep Medicine*, 2nd ed., ed. Meir H. Kryger, Alon Y. Avidan, and Richard B. Berry, 353–56. Philadelphia: Elsevier/Saunders, 2014.

Nandalike, K., et al. "Screening for Sleep-Disordered Breathing and Excessive Daytime Sleepiness in Adolescent Girls with Polycystic Ovarian Syndrome." *Journal of Pediatrics* 159, no. 4 (2011): 591–96. doi:10.1016/j.jpeds.2011.04.027.

Shah, D., and S. Rasool. "Polycystic Ovary Syndrome and Metabolic Syndrome: The Worrisome Twosome?" *Climacteric* 19, no. 1 (2016): 7–16. doi:10.3109/136 97137.2015.1116505.

Shechter, A., et al. "Nocturnal Polysomnographic Sleep Across the Menstrual Cycle in Premenstrual Dysphoric Disorder." *Sleep Medicine* 13, no. 8 (2012): 1071–78. doi:10.1016/j.sleep.2012.05.012.

Takeda, T., et al. "Fish Consumption and Premenstrual Syndrome and Dysphoric Disorder in Japanese Collegiate Athletes." *Journal of Pediatric and Adolescent Gynecology* 29, no. 4 (2016): 386–89.

Van Reen, E., and J. Kiesner. "Individual Differences in Self-Reported Difficulty Sleeping Across the Menstrual Cycle." *Archives of Women's Mental Health* 19, no. 4 (2016): 599–608. doi:10.1007/s00737-016-0621-9.

CHAPTER 4: PREGNANCY AND POSTPARTUM

Balserak, B., and K. Lee. "Sleep and Sleep Disorders Associated with Pregnancy." In *Principles and Practice of Sleep Medicine*, 6th ed., ed. Meir H. Kryger, Thomas Roth, and William C. Dement, 1525–39. Philadelphia: Elsevier, 2017.

Bin, Y. S., et al. "Population-Based Study of Sleep Apnea in Pregnancy and Maternal and Infant Outcomes." *Journal of Clinical Sleep Medicine* 12, no. 6 (2016): 871–77.

Facco, F., et al. "Sleep Disordered Breathing in Pregnancy." In *Principles and Practice of Sleep Medicine*, 6th ed., ed. Meir H. Kryger, Thomas Roth, and William C. Dement, 1540–46. Philadelphia: Elsevier, 2017.

Gupta, R., et al. "Restless Legs Syndrome and Pregnancy: Prevalence, Possible Pathophysiological Mechanisms and Treatment." *Acta Neurologica Scandinavica* 133, no. 5 (2016): 320–29. doi:10.1111/ane.12520.

Krawczak, E. M., et al. "Do Changes in Subjective Sleep and Biological Rhythms Predict Worsening in Postpartum Depressive Symptoms? A Prospective Study Across the Perinatal Period." *Archives of Women's Mental Health* 19, no. 4 (2016): 591–98. doi:10.1007/s00737-016-0612-x.

National Sleep Foundation. "Summary of Findings: 2007 NSF Sleep in America Poll." https://sleepfoundation.org/sites/default/files/Summary_Of_Findings %20-%20FINAL.pdf (accessed August 25, 2016).

Pamidi, S., et al. 2016. "Maternal Sleep-Disordered Breathing and the Risk of Delivering Small for Gestational Age Infants: A Prospective Cohort Study." *Thorax* 71, no. 8 (2016): 719–25. doi:10.1136/thoraxjnl-2015-208038.

Paszkowski, M., et al. "Selected Non-Somatic Risk Factors for Pregnancy Loss in Patients with Abnormal Early Pregnancy." *Annals of Agricultural and Environmental Medicine* 23, no. 1 (2016): 153–56. doi:10.5604/12321966.1196872.

Sharma, S. K., et al. "Sleep Disorders in Pregnancy and Their Association with Pregnancy Outcomes: A Prospective Observational Study." *Sleep and Breathing* 20, no. 1 (2016): 87–93. doi:10.1007/s11325-015-1188-9.

Stremler, R., et al. "Postpartum Period and Early Motherhood." In *Principles and Practice of Sleep Medicine*, 6th ed., ed. Meir H. Kryger, Thomas Roth, and William C. Dement, 1547–52. Philadelphia: Elsevier, 2017.

Whitehead, C., et al. "Treatment of Early-Onset Preeclampsia with Continuous Positive Airway Pressure." *Obstetrics & Gynecology* 125, no. 5 (2015): 1106–9. doi:10.1097/AOG.0000000000000508.

Yuen, Kin M. "Pregnancy and Postpartum." In *Atlas of Clinical Sleep Medicine*, 2nd ed., ed. Meir H. Kryger, Alon Y. Avidan, and Richard B. Berry, 356–59. Philadelphia: Elsevier/Saunders, 2014.

CHAPTER 5: WHEN SEX HORMONE LEVELS DECREASE

Avis, N. E., et al. "Duration of Menopausal Vasomotor Symptoms over the Menopause Transition." *Journal of the American Medical Association Internal Medicine* 175, no. 4 (2015): 531–39. doi:10.1001/jamainternmed.2014.8063.

Baker, F. C., A. R. Willoughby, S. A. Sassoon, I. M. Colrain, and M. de Zambotti. "Insomnia in Women Approaching Menopause: Beyond Perception." *Psychoneuroendocrinology* 60 (2015): 96–104. doi:10.1016/j.psyneuen.2015.06.005.

Baker, Fiona, et al. "Sleep and Menopause." In *Principles and Practice of Sleep Medicine*, 6th ed., ed. Meir H. Kryger, Thomas Roth, and William C. Dement, 1553–63. Philadelphia: Elsevier, 2017.

Brody, Jane. "Tackling Menopause's Side Effects." *New York Times*, February 10, 2014. http://well.blogs.nytimes.com/2014/02/10/tackling-menopauses-side-effects/.

Bruyneel, M. "Sleep Disturbances in Menopausal Women: Aetiology and Practical Aspects." *Maturitas* 81, no. 3 (2015): 406–9. doi:10.1016/j.maturitas.2015 .04.017.

Cumming, G. P., et al. "The Need to Do Better—Are We Still Letting Our

Patients Down and at What Cost?" *Post Reproductive Health* 21, no. 2 (2015): 56–62. doi:10.1177/2053369115586122.

Gentry-Maharaj, A., et al. "Use and Perceived Efficacy of Complementary and Alternative Medicines After Discontinuation of Hormone Therapy: A Nested United Kingdom Collaborative Trial of Ovarian Cancer Screening Cohort Study." *Menopause* 22, no. 4 (2015): 384–90. doi:10.1097/GME.0000000000000330.

Grant, M. D., A. Marbella, A. T. Wang, E. Pines, et al. *Menopausal Symptoms: Comparative Effectiveness of Therapies.* Rockville, Md.: Agency for Healthcare Research and Quality (US), 2015.

Jakiel, G., et al. "Andropause—State of the Art 2015 and Review of Selected Aspects." *Menopause Review* 14, no. 1 (2015): 1–6. doi:10.5114/pm.2015.49998.

Jiang, Bei, Fredi Kronenberg, Paiboon Nuntanakorn, Min-Hua Qiu, and Edward J. Kennelly. "Evaluation of the Botanical Authenticity and Phytochemical Profile of Black Cohosh Products by High-Performance Liquid Chromatography with Selected Ion Monitoring Liquid Chromatography–Mass Spectrometry." *Journal of Agricultural and Food Chemistry* 54, no. 9 (2006): 3242–53. doi:10.1021/jf0606149.

Kravitz, H. M., et al. "Sleep Difficulty in Women at Midlife: A Community Survey of Sleep and the Menopausal Transition." *Menopause* 10, no. 1 (2003): 19–28.

Minkin, M. J., et al. "Prevalence of Postmenopausal Symptoms in North America and Europe." *Menopause* 22, no. 11 (2015): 1231–38. doi:10.1097/GME.0000000000000464.

National Sleep Foundation. "Summary of Findings: 2007 NSF Sleep in America Poll." https://sleepfoundation.org/sites/default/files/Summary_Of_Findings%20-%20FINAL.pdf (accessed August 25, 2016).

CHAPTER 6: HOW TO IDENTIFY A SLEEP PROBLEM

"Berlin Questionnaire." British Snoring & Sleep Apnoea Association. http://www.britishsnoring.co.uk/berlin_questionnaire.php (accessed August 25, 2016).

"Epworth Sleepiness Scale." British Snoring & Sleep Apnoea Association. http://www.britishsnoring.co.uk/sleep_apnoea/epworth_sleepiness_scale.php (Accessed August 25, 2016).

"Narcolepsy: Self Evaluation." Harvard Medical School. http://healthysleep.med.harvard.edu/narcolepsy/diagnosing-narcolepsy/narcolepsy-self-evaluation (accessed August 25, 2016).

"Narcolepsy: Ullanlinna Narcolepsy Scale." Harvard Medical School. http://healthysleep.med.harvard.edu/file/56 (accessed August 25, 2016).

"STOPBang Questionnaire." British Snoring & Sleep Apnoea Association. http://www.britishsnoring.co.uk/stop_bang_questionnaire.php (accessed August 25, 2016).

"Two Week Sleep Diary." American Academy of Sleep Medicine. http://
    yoursleep.aasmnet.org/pdf/sleepdiary.pdf (accessed August 25, 2016).
Vaughn, Bradley V., and O'Neill F. D'Cruz. "Cardinal Manifestations of Sleep
    Disorders." In *Principles and Practice of Sleep Medicine*, 6th ed., ed. Meir H.
    Kryger, Thomas Roth, and William C. Dement, 573–87. Philadelphia: Else-
    vier, 2017.

CHAPTER 7: SECONDHAND SLEEP PROBLEMS

AAP Task Force on Sudden Infant Death Syndrome. "SIDS and Other Sleep-
    Related Infant Deaths: Updated 2016 Recommendations for a Safe Infant
    Sleeping Environment." *Pediatrics* 138, no. 5 (2016): e20162938. http://pediat
    rics.aappublications.org/content/pediatrics/early/2016/10/20/peds.2016-2938
    .full.pdf.
Fadini, C. C., et al. "Influence of Sleep Disorders on the Behavior of Individuals
    with Autism Spectrum Disorder." *Frontiers of Human Neuroscience* 9 (2015):
    347. doi:10.3389/fnhum.2015.00347.
Ferber, Richard. *Solve Your Child's Sleep Problems.* New, revised, and expanded
    ed. New York: Fireside Books, 2006.
Gradisar, M., et al. "Behavioral Interventions for Infant Sleep Problems: A Ran-
    domized Controlled Trial." *Pediatrics* 137, no. 6 (2016): 1–10; e20151486. doi:
    10.1542/peds.2015-1486.
Kabeshita, Y., et al. "Sleep Disturbances are Key Symptoms of Very Early Stage
    Alzheimer Disease with Behavioral and Psychological Symptoms: A Japan
    Multi-Center Cross-Sectional Study (J-BIRD)." *International Journal of Geri-
    atric Psychiatry* (2016). doi:10.1002/gps.4470.
Lai, A. Y., et al. 2015. "A Pathway Underlying the Impact of CPAP Adherence
    on Intimate Relationship with Bed Partner in Men with Obstructive Sleep
    Apnea." *Sleep and Breathing* 20, no. 2 (2015): 543–51. doi:10.1007/s11325-015
    -1235-6.
Liguori, C., et al. "Rapid Eye Movement Sleep Disruption and Sleep Fragmen-
    tation Are Associated with Increased Orexin-A Cerebrospinal-Fluid Levels in
    Mild Cognitive Impairment Due to Alzheimer's Disease." *Neurobiology of
    Aging* 40 (2016): 120–26. doi:10.1016/j.neurobiolaging.2016.01.007.
McCleery, J., et al. "Pharmacotherapies for Sleep Disturbances in Alzheimer's
    Disease." *Cochrane Database Systemic Reviews* 3 (2014): CD009178. doi:10
    .1002/14651858.CD009178.pub2.
Mindell, Jodi A. *Sleeping Through the Night: How Infants, Toddlers, and Their Par-
    ents Can Get a Good Night's Sleep.* Rev. ed. New York: William Morrow, 2005.
Mindell, Jodi A., and Judith A. Owens. *A Clinical Guide to Pediatric Sleep:*

Diagnosis and Management of Sleep Problems, 3rd ed. Philadelphia: Wolters Kluwer/Lippincott Williams and Wilkins, 2015.

Osorio, R. S., et al. "Sleep-Disordered Breathing Advances Cognitive Decline in the Elderly." *Neurology* 84, no. 19 (2015): 1964–71. doi:10.1212/WNL.0000000000001566.

Pediatric Sleep Council. Interactive website on babies and sleep: https://www.babysleep.com.

"Suffocation Deaths Associated with Use of Infant Sleep Positioners—United States, 1997–2011." Centers for Disease Control and Prevention, November 23, 2012. http://www.cdc.gov/mmwr/preview/mmwrhtml/mm6146a1.htm (accessed August 25, 2016).

Weissbluth, Marc. *Healthy Sleep Habits, Happy Child: A Step-by-Step Program for a Good Night's Sleep*, 4th ed. New York: Ballantine Books, 2015.

CHAPTER 8: RESETTING THE BODY CLOCK

Abbott, S. M., et al. "Circadian Rhythm Sleep-Wake Disorders." *Psychiatry Clinics of North America* 38, no. 4 (2015): 805–23. doi:10.1016/j.psc.2015.07.012.

Crowley, S. J., and Eastman, C. I. "Phase Advancing Human Circadian Rhythms with Morning Bright Light, Afternoon Melatonin, and Gradually Shifted Sleep: Can We Reduce Morning Bright-Light Duration?" *Sleep Medicine* 16, no. 2 (2015): 288–97. doi:10.1016/j.sleep.2014.12.004.

Feillet, C., et al. "Coupling Between the Circadian Clock and Cell Cycle Oscillators: Implication for Healthy Cells and Malignant Growth." *Frontiers of Neurology and Neuroscience* 6 (2015): 96. doi:10.3389/fneur.2015.00096.

Preckel, F., et al. "Morningness-Eveningness and Educational Outcomes: The Lark Has an Advantage over the Owl at High School." *British Journal of Educational Psychology* 83 (Pt 1) (2013): 114–34. doi:10.1111/j.2044-8279.2011.02059.x.

Puram, R. V., et al. "Core Circadian Clock Genes Regulate Leukemia Stem Cells in AML." *Cell* 165, no. 2 (2016): 303–16. doi:10.1016/j.cell.2016.03.015.

Reid, K. J., and S. M. Abbott. "Jet Lag and Shift Work Disorder." *Sleep Medicine Clinics* 10, no. 4 (2015): 523–35. doi:10.1016/j.jsmc.2015.08.006.

Thacher, P. V. "University Students and 'the All Nighter': Correlates and Patterns of Students' Engagement in a Single Night of Total Sleep Deprivation." *Behavioral Sleep Medicine* 6, no. 1 (2008): 16–31. doi:10.1080/15402000701796114.

Tortorolo, F., et al. "Is Melatonin Useful for Jet Lag?" *Medwave* 15 (Suppl. 3) (2015): e6343. doi:10.5867/medwave.2015.6343.

Videnovic, A., and P. C. Zee,. "Consequences of Circadian Disruption on Neurologic Health." *Sleep Medicine Clinics* 10, no. 4 (2015): 469–80. doi:10.1016/j.jsmc.2015.08.004.

Zee, P. C. "Circadian Clocks: Implication for Health and Disease." *Sleep Medi-cine Clinics* 10, no. 4 (2015): xiii. doi:10.1016/j.jsmc.2015.09.002.

CHAPTER 9: A WORLD THAT NEVER SLEEPS

Abbott, S. M., et al. "Circadian Rhythm Sleep-Wake Disorders." *Psychiatry Clin-ics of North America* 38, no. 4 (2015): 805–23. doi:10.1016/j.psc.2015.07.012.
Cordina-Duverger, E., Y. Koudou, T. Truong, P. Arveux, et al. "Night Work and Breast Cancer Risk Defined by Human Epidermal Growth Factor Receptor-2 (HER2) and Hormone Receptor Status: A Population-Based Case-Control Study in France." *Chronobiology International* 33, no. 6 (2016): 783–87. doi: 10.3109/07420528.2016.1167709.
Davis, S., et al. "Night Shift Work, Light at Night, and Risk of Breast Cancer." *Journal of the National Cancer Institute* 93, no. 20 (2001): 1557–62.
Feillet, C., et al. "Coupling Between the Circadian Clock and Cell Cycle Os-cillators: Implication for Healthy Cells and Malignant Growth." *Frontiers of Neurology and Neuroscience* 6 (2015): 96. doi:10.3389/fneur.2015.00096.
Huffington, Arianna. *The Sleep Revolution: Transforming Your Life, One Night at a Time.* New York: Harmony, 2016.
Huffmyer, J. L., et al. "Driving Performance of Residents After Six Consec-utive Overnight Work Shifts." *Anesthesiology* 124, no. 6 (2016): 1396–403. doi:10.1097/ALN.0000000000001104.
Lin, Y. C., et al. "Effect of Rotating Shift Work on Childbearing and Birth Weight: A Study of Women Working in a Semiconductor Manufacturing Factory." *World Journal of Pediatrics* 7, no. 2 (2011): 129–35. doi:10.1007/s12519 -011-0265-9.
Logan, R. W., et al. "Chronic Shift-Lag Alters the Circadian Clock of NK Cells and Promotes Lung Cancer Growth in Rats." *Journal of Immunology* 188, no. 6 (2012): 2583–91. doi:10.4049/jimmunol.1102715.
Rabstein, S., et al. "Polymorphisms in Circadian Genes, Night Work and Breast Cancer: Results from the GENICA Study." *Chronobiology International* 31, no. 10 (2014): 1115–22. doi:10.3109/07420528.2014.957301.
Reid, K. J., and S. M. Abbott. "Jet Lag and Shift Work Disorder." *Sleep Medicine Clinics* 10, no. 4 (2015): 523–35. doi:10.1016/j.jsmc.2015.08.006.

CHAPTER 10: INSOMNIA

"2015 Sleep in America Poll Finds Pain a Significant Challenge When It Comes to Americans' Sleep," March 2, 2015, National Sleep Foundation. https://

sleepfoundation.org/media-center/press-release/2015-sleep-america-poll (accessed August 25, 2016).

Almondes, K. M., et al. "Insomnia and Risk of Dementia in Older Adults: Systematic Review and Meta-Analysis." *Journal of Psychiatry Research* 77 (2016): 109–15. doi:10.1016/j.jpsychires.2016.02.021.

DiNapoli, E. A., et al. "Sedative Hypnotic Use Among Veterans with a Newly Reported Mental Health Disorder." *International Psychogeriatrics* 28, no. 8 (2016): 1391–98. doi:10.1017/S1041610216000521.

Edinger, Jack D., Meir H. Kryger, and Thomas Roth. "Insomnia." In *Atlas of Clinical Sleep Medicine*, 2nd ed., ed. Meir H. Kryger, Alon Y. Avidan, and Richard B. Berry, 148–58. Philadelphia: Elsevier/Saunders, 2014.

Frackt, Austin. "How an Insomnia Therapy Can Help with Other Illnesses." *New York Times*, July 13, 2015. http://www.nytimes.com/2015/07/14/upshot/how-an-insomnia-therapy-can-help-with-other-illnesses.html.

Grandner, M. A., et al. "Sleep Symptoms, Race/Ethnicity, and Socioeconomic Position." *Journal of Clinical Sleep Medicine* 9, no. 9 (2013): 897–905, 905A–905D. doi:10.5664/jcsm.2990.

Lichstein, K., et al. "Insomnia: Epidemiology and Risk Factors." In *Principles and Practice of Sleep Medicine*, 6th ed., ed. Meir H. Kryger, Thomas Roth, and William C. Dement, 761–68. Philadelphia: Elsevier, 2017.

Perlis, Michael L., Michael T. Smith, and Wilfred R. Pigeon. "Etiology and Pathophysiology of Insomnia." In *Principles and Practice of Sleep Medicine*, 6th ed., ed. Meir H. Kryger, Thomas Roth, and William C. Dement, 769–84. Philadelphia: Elsevier, 2017.

CHAPTER 11: RESTLESS LEGS SYNDROME

Allen, Richard P., Rachel E. Salas, and Charlene Gamaldo. "Movement Disorders in Sleep." In *Atlas of Clinical Sleep Medicine*, 2nd ed., ed. Meir H. Kryger, Alon Y. Avidan, and Richard B. Berry, 174–94. Philadelphia: Elsevier/Saunders, 2014.

Cassel, W., et al. "Significant Association Between Systolic and Diastolic Blood Pressure Elevations and Periodic Limb Movements in Patients with Idiopathic Restless Legs Syndrome." *Sleep Medicine* 17 (2016): 109–20. doi:10.1016/j.sleep.2014.12.019.

Gupta, R., et al. "Restless Legs Syndrome and Pregnancy: Prevalence, Possible Pathophysiological Mechanisms and Treatment." *Acta Neurologica Scandinavica* 133, no. 5 (2016): 320–29. doi:10.1111/ane.12520.

Harashima, S., et al. "Restless Legs Syndrome in Patients with Type 2 Diabetes:

Effectiveness of Pramipexole Therapy." *BMJ Supportive & Palliative Care* 6, no 1 (2016): 89–93. doi:10.1136/bmjspcare-2014-000691.

Moccia, M., et al. "A Four-Year Longitudinal Study on Restless Legs Syndrome in Parkinson Disease." *Sleep* 39, no. 2 (2016): 405–12. doi:10.5665/sleep.5452.

Oyieng'o, D. O., et al. "Restless Legs Symptoms and Pregnancy and Neonatal Outcomes." *Clinical Therapeutics* 38, no. 2 (2016): 256–64. doi:10.1016/j .clinthera.2015.11.021.

Pratt, D. P. "Restless Legs Syndrome/Willis-Ekbom Disease and Periodic Limb Movements: A Comprehensive Review of Epidemiology, Pathophysiology, Diagnosis and Treatment Considerations." *Current Rheumatology Reviews* 12, no. 2 (2016): 91–112.

Rinaldi, F., et al. "Treatment Options in Intractable Restless Legs Syndrome/ Willis-Ekbom Disease (RLS/WED)." *Current Treatment Options in Neurology* 18, no. 2 (2016): 7. doi:10.1007/s11940-015-0390-1.

Trenkwalder, Claudia, et al. "Restless Legs Syndrome Associated with Major Diseases: A Systematic Review and New Concept." *Neurology* 86, no. 14 (2016): 1336–43. doi:10.1212/WNL.0000000000002542.

Xue, R., et al. "An Epidemiologic Study of Restless Legs Syndrome Among Chinese Children and Adolescents." *Neurological Sciences* 36, no. 6 (2015): 971–76. doi:10.1007/s10072-015-2206-1.

CHAPTER 12: SLEEP APNEA

"Berlin Questionnaire." British Snoring & Sleep Apnoea Association. http:// www.britishsnoring.co.uk/berlin_questionnaire.php (accessed May 2016).

"BMI Calculator for Adults." National Heart, Lung, and Blood Institute. http:// www.nhlbi.nih.gov/health/educational/lose_wt/BMI/bmicalc.htm (accessed August 25, 2016).

"BMI Percentile Calculator for Child and Teen: English Version." Centers for Disease Control and Prevention. https://nccd.cdc.gov/dnpabmi/calculator .aspx (accessed August 25, 2016).

"Body Mass Index." Centers for Disease Control and Prevention. http://www.cdc .gov/healthyweight/assessing/bmi/index.html (accessed August 25, 2016).

Global BMI Mortality Collaboration. "Body-Mass Index and All-Cause Mortality: Individual-Participant-Data Meta-Analysis of 239 Prospective Studies in Four Continents." *The Lancet* 388, no. 10046 (2016): 776–86. doi: 10.1016/ S0140-6736(16)30175-1.

Hayes, Don, Jr., Mark Splaingard, and Meir H. Kryger. "Sleep Apnea in the Adolescent and Adult." In *Atlas of Clinical Sleep Medicine*, 2nd ed., ed. Meir H.

Kryger, Alon Y. Avidan, and Richard B. Berry, 269–99. Philadelphia: Elsevier/ Saunders, 2014.

Kryger, Meir H., et al. "The Sleep Deprivation Syndrome of the Obese Patient: A Problem of Periodic Nocturnal Upper Airway Obstruction." *American Journal of Medicine* 56, no. 4 (1974): 530–39.

Mitler, M. M., et al. "Bedtime Ethanol Increases Resistance of Upper Airways and Produces Sleep Apneas in Asymptomatic Snorers." *Alcoholism: Clinical and Experimental Research* 12, no. 6 (1988): 801–5.

NCD Risk Factor Collaboration (NCD-RisC). "Trends in Adult Body-Mass Index in 200 Countries from 1975 to 2014: A Pooled Analysis of 1698 Population-Based Measurement Studies with 19.2 Million Participants." *The Lancet* 387, no. 10026 (2016): 1377–96. doi: 10.1016/S0140-6736(16)30054-X.

Sheldon, Stephen H. "Obstructive Sleep Apnea in Children." In *Atlas of Clinical Sleep Medicine*, 2nd ed., ed. Meir H. Kryger, Alon Y. Avidan, and Richard B. Berry, 299–307. Philadelphia: Elsevier/Saunders, 2014.

Sotos, J. G. "Taft and Pickwick: Sleep Apnea in the White House." *Chest* 124, no. 3 (2003): 1133–42.

"STOPBang Questionnaire." British Snoring & Sleep Apnoea Association. http:// www.britishsnoring.co.uk/stop_bang_questionnaire.php (accessed May 2016).

Sullivan, Colin E., Michael Berthon-Jones, Faiq G. Issa, and Lorraine Eves. "Reversal of Obstructive Sleep Apnoea by Continuous Positive Airway Pressure Applied Through the Nares." *The Lancet* 317, no. 8225 (1981): 862–65.

CHAPTER 13: NARCOLEPSY

Appold, Karen. "Narcolepsy's New Names." *Sleep Review Magazine.* Last modified April 16, 2014. http://www.sleepreviewmag.com/2014/04/narcolepsys-new -names/.

Arango, M. T., et al. "Is Narcolepsy a Classical Autoimmune Disease?" *Pharmacological Research* 92 (2015): 6–12. doi:10.1016/j.phrs.2014.10.005.

Broughton, R. J., et al. "Randomized, Double-Blind, Placebo-Controlled Cross-over Trial of Modafinil in the Treatment of Excessive Daytime Sleepiness in Narcolepsy." *Neurology* 49, no. 2 (1997): 444–51.

Chow, Matthew, and Michelle Cao. "The Hypocretin/Orexin System in Sleep Disorders: Preclinical Insights and Clinical Progress." *Nature and Science of Sleep* 8 (2016): 81–86. doi:10.2147/NSS.S76711.

Fraigne, J. J., et al. "REM Sleep at Its Core — Circuits, Neurotransmitters, and Pathophysiology." *Frontiers of Neurology and Neuroscience* 6 (2015): 123. doi:10.3389/fneur.2015.00123.

Gavrilov, Y. V., et al. "Disrupted Sleep in Narcolepsy: Exploring the Integrity of Galanin Neurons in the Ventrolateral Preoptic Area." *Sleep* 39, no. 5 (May 1 2016): 1059–62. doi:10.5665/sleep.5754.

Khan, Z., and L. M. Trotti. "Central Disorders of Hypersomnolence: Focus on the Narcolepsies and Idiopathic Hypersomnia." *Chest* 148, no. 1 (2015): 262–73. doi:10.1378/chest.14-1304.

Kryger, Meir H., et al. "Diagnoses Received by Narcolepsy Patients in the Year Prior to Diagnosis by a Sleep Specialist." *Sleep* 25, no. 1 (2002): 36–41.

Montplaisir, J., et al. "Risk of Narcolepsy Associated with Inactivated Adjuvanted (AS03) A/H1N1 (2009) Pandemic Influenza Vaccine in Quebec." *PLOS One* 9, no. 9 (2014): e108489. doi:10.1371/journal.pone.0108489.

"Narcolepsy: Self-Evaluation." Harvard Medical School. Last modified July 22, 2013. http://healthysleep.med.harvard.edu/narcolepsy/diagnosing-narcolepsy/narcolepsy-self-evaluation (accessed August 25, 2016).

"Narcolepsy: Ullanlinna Narcolepsy Scale." Harvard Medical School. http://healthysleep.med.harvard.edu/file/56 (accessed August 25, 2016).

Scammell, T. E. "Narcolepsy." *New England Journal of Medicine* 373, no. 27 (2015): 2654–62. doi:10.1056/NEJMra1500587.

CHAPTER 14: FEAR OF SLEEPING AND
OTHER UNUSUAL AILMENTS

Carrillo-Solano, M., et al. "Sleepiness in Sleepwalking and Sleep Terrors: A Higher Sleep Pressure?" *Sleep Medicine* (January 4, 2016). doi:10.1016/j.sleep.2015.11.020.

Cochen De Cock, V. "Sleepwalking." *Current Treatment Options in Neurology* 18, no. 2 (2016): 6. doi:10.1007/s11940-015-0388-8.

Horvath, András, Anikó Papp, and Anna Szűcs. "Progress in Elucidating the Pathophysiological Basis of Nonrapid Eye Movement Parasomnias: Not Yet Informing Therapeutic Strategies." *Nature and Science of Sleep* 8 (2016): 73–79. doi:10.2147/NSS.S71513.

Howell, Michael J., and Carlos H. Schenck. "Parasomnias." In *Atlas of Clinical Sleep Medicine*, 2nd ed., ed. Meir H. Kryger, Alon Y. Avidan, and Richard B. Berry, 237–53. Philadelphia: Elsevier/Saunders, 2014.

Januszko, P., et al. "Sleepwalking Episodes Are Preceded by Arousal-Related Activation in the Cingulate Motor Area: EEG Current Density Imaging." *Clinical Neurophysiology* 127, no. 1 (2016): 530–36. doi:10.1016/j.clinph.2015.01.014.

Jiang, H., et al. "RBD and Neurodegenerative Diseases." *Molecular Neurobiology* (2016). doi:10.1007/s12035-016-9831-4.

Kryger, Meir H. "PTSD: Not Just in Soldiers, and Not Just in Men." *Huffington Post*, May 18, 2015, updated May 18, 2016. http://www.huffingtonpost.com/society-for-womens-health-research/ptsd-not-just-in-soldiers-and-not-just-in-men_b_7292154.html.

Kryger, Meir H. "The Tigers Come at Night: REM Sleep Disorder." *Psychology Today*, October 29, 2013. https://www.psychologytoday.com/blog/sleep-and-be-well/201310/the-tigers-come-night.

CHAPTER 15: MEDICAL CONDITIONS THAT AFFECT SLEEP

Al Mawed, S., and M. Unruh. "Diabetic Kidney Disease and Obstructive Sleep Apnea: A New Frontier?" *Current Opinions in Pulmonary Medicine* 22, no. 1 (2016): 80–88. doi:10.1097/MCP.0000000000000230.

Arnaldi, D., et al. "Does Postural Rigidity Decrease During REM Sleep Without Atonia in Parkinson Disease?" *Journal of Clinical Sleep Medicine* 12, no. 6 (2016): 839–47.

Chwiszczuk, L., et al. "Higher Frequency and Complexity of Sleep Disturbances in Dementia with Lewy Bodies as Compared to Alzheimer's Disease." *Neurodegenerative Diseases* 16, no. 3–4 (2016): 152–60. doi:10.1159/000439252.

Fyfe, I. "Parkinson Disease: Sleep Disorder Deficits Suggest Signature for Early Parkinson Disease." *Nature Reviews Neurology* 12, no. 1 (2016): 3. doi:10.1038/nrneurol.2015.232.

Goichot, B., et al. "Clinical Presentation of Hyperthyroidism in a Large Representative Sample of Outpatients in France: Relationships with Age, Aetiology and Hormonal Parameters." *Clinical Endocrinology (Oxf)* 84, no. 3 (2016): 445–51. doi:10.1111/cen.12816.

Pearse, S. G., and M. R. Cowie. "Sleep-Disordered Breathing in Heart Failure." *European Journal of Heart Failure* 18, no. 4 (2016): 353–61. doi:10.1002/ejhf.492.

Roth, T., P. Bhadra-Brown, V. W. Pitman, and E. M Resnick. "Pregabalin Improves Fibromyalgia-Related Sleep Disturbance." *Clinical Journal of Pain* 32, no. 4 (2016): 308–12. doi:10.1097/AJP.0000000000000262.

Roth, T., P. Bhadra-Brown, V. W. Pitman, T. A. Roehrs, and E. M Resnick. "Characteristics of Disturbed Sleep in Patients with Fibromyalgia Compared with Insomnia or with Pain-Free Volunteers." *Clinical Journal of Pain* 32, no. 4 (2016): 302–7. doi:10.1097/AJP.0000000000000261.

Sridhar, G. R., et al. "Sleep in Thyrotoxicosis." *Indian Journal of Endocrinology and Metabolism* 15, no. 1 (2011): 23–26. doi:10.4103/2230-8210.77578.

Zoetmulder, M., et al. "Increased Motor Activity During REM Sleep Is Linked

with Dopamine Function in Idiopathic REM Sleep Behaviour Disorder
and Parkinson Disease." *Journal of Clinical Sleep Medicine* 12, no. 6 (2016):
895–903.

CHAPTER 16: PSYCHIATRIC DISORDERS THAT AFFECT SLEEP

Annamalai, A., et al. "High Rates of Obstructive Sleep Apnea Symptoms
Among Patients with Schizophrenia." *Psychosomatics* 56, no. 1 (2016): 59–66.
doi:10.1016/j.psym.2014.02.009.

Baglioni, C., et al. "Insomnia as a Predictor of Depression: A Meta-Analytic
Evaluation of Longitudinal Epidemiological Studies." *Journal of Affective
Disorders* 135, nos. 1–3 (2011): 10–19. doi:10.1016/j.jad.2011.01.011.

"Bipolar Disorder." National Institute of Mental Health. https://www.nimh.nih
.gov/health/topics/bipolar-disorder/ (accessed August 30, 2016).

Chan, M. S., et al. "Sleep in Schizophrenia: A Systematic Review and Meta-
Analysis of Polysomnographic Findings in Case-Control Studies." *Sleep Medicine Reviews* (2016). doi:10.1016/j.smrv.2016.03.001.

Chang, P. P., et al. "Insomnia in Young Men and Subsequent Depression: The
Johns Hopkins Precursors Study." *American Journal of Epidemiology* 146, no.
2 (1997): 105–14.

"Depression." National Institute of Mental Health. https://www.nimh.nih.gov/
health/topics/depression/ (accessed August 30, 2016).

Gupta, M. A., and F. C. Simpson. "Obstructive Sleep Apnea and Psychiatric
Disorders: A Systematic Review." *Journal of Clinical Sleep Medicine* 11, no. 2
(2015): 165–75. doi:10.5664/jcsm.4466.

Kripke, Daniel F., et al. "Photoperiodic and Circadian Bifurcation Theories
of Depression and Mania." *F1000Research* 4 (2015): 107. doi:10.12688/f1000
research.6444.1.

Kryger, Meir H. "Collateral Damage: Nightmares in PTSD." *Psychology Today*,
March 1, 2013. https://www.psychologytoday.com/blog/sleep-and-be-well/
201303/collateral-damage.

Kryger, Meir H. "PTSD: Not Just in Soldiers, and Not Just in Men." *Huffington
Post*. May 18, 2015, updated May 18, 2016. http://www.huffingtonpost.com/
society-for-womens-health-research/ptsd-not-just-in-soldiers-and-not-just-in
-men_b_7292154.html.

Linde, K., L. Kriston, G. Rücker, S. Jamil, et al. "Efficacy and Acceptability
of Pharmacological Treatments for Depressive Disorders in Primary Care:
Systematic Review and Network Meta-analysis." *Annals of Family Medicine*
13, no. 1 (2015): 69–79. doi:10.1186/s12875-015-0314-x.

Lipinska, G., et al. "Pharmacology for Sleep Disturbance in PTSD." *Human Psychopharmacology* 31, no. 2 (2016): 156–63. doi:10.1002/hup.2522.

Remes, O., C. Brayne, Rianne van der Linde, and Louise Lafortune. "A Systematic Review of Reviews on the Prevalence of Anxiety Disorders in Adult Populations." *Brain and* Behavior 6, no. 7 (2016): e00497. https://www.ncbi.nlm.nih.gov/pmc/articles/PMC4951626/.

Shepertycky, M. R., et al. "Differences Between Men and Women in the Clinical Presentation of Patients Diagnosed with Obstructive Sleep Apnea Syndrome." *Sleep* 28, no. 3 (2005): 309–14.

Tseng, P. T., et al. "Light Therapy in the Treatment of Patients with Bipolar Depression: A Meta-Analytic Study." *European Journal of Neuropsychopharmacology* 26, no. 6 (2016): 1037–47. doi:10.1016/j.euroneuro.2016.03.001.

Yildiz, M., et al. "State of the Art Psychopharmacological Treatment Options in Seasonal Affective Disorder." *Psychiatria Danubina* 28, no. 1 (2016): 25–29.

Yokoyama, E., Y. Kaneita, Y. Saito, M. Uchiyama, et al. "Association Between Depression and Insomnia Subtypes: A Longitudinal Study on the Elderly in Japan." *Sleep* 33, no. 12 (2010): 1693–702.

CHAPTER 17: MEDICATIONS THAT CONTRIBUTE TO SLEEP DISORDERS

"Caffeine Content of Products." Caffeineinformer.com. http://www.caffeineinformer.com/the-caffeine-database (accessed August 25, 2016).

"Club Drugs." Last modified December 2012. National Institute on Drug Abuse. https://www.drugabuse.gov/drugs-abuse/club-drugs (accessed August 25, 2016).

Kidwell, K. M., et al. "Stimulant Medications and Sleep for Youth with ADHD: A Meta-analysis." *Pediatrics* 136, no. 6 (2015): 1144–53. doi:10.1542/peds.2015-1708.

National Center for Health Statistics. *Health, United States, 2015: With Special Feature on Racial and Ethnic Health Disparities.* Table 79: "Prescription Drug Use in the Past 30 Days, by Sex, Race and Hispanic Origin, and Age: United States, Selected Years 1988–1994 Through 2009–2012." Hyattsville, Md.: U.S. GPO, 2016. http://www.cdc.gov/nchs/data/hus/hus15.pdf#079.

Scheer, F. A., et al. "Repeated Melatonin Supplementation Improves Sleep in Hypertensive Patients Treated with Beta-Blockers: A Randomized Controlled Trial." *Sleep* 35, no. 10 (2012): 1395–402. doi:10.5665/sleep.2122.

Schweitzer, Paula K., and A. Rondozzo. "Drugs That Disturb Sleep and Wakefulness." In *Principles and Practice of Sleep Medicine*, 6th ed., ed. Meir

H. Kryger, Thomas Roth, and William C. Dement, 480–98. Philadelphia: Elsevier, 2017.

Takada, M., et al. "Association of Statin Use with Sleep Disturbances: Data Mining of a Spontaneous Reporting Database and a Prescription Database." *Drug Safety* 37, no. 6 (2014): 421–31. doi:10.1007/s40264-014-0163-x.

CHAPTER 18: AT THE SLEEP CLINIC

Cooksey, J. A., and J. S. Balachandran. "Portable Monitoring for the Diagnosis of OSA." *Chest* 149, no. 4 (2016): 1074–81. doi:10.1378/chest.15-1076.

"Find a Sleep Facility Near You [USA]." American Academy of Sleep Medicine. http://www.sleepeducation.org/find-a-facility.

Johnson, K. G., and D. C. Johnson. "Treatment of Sleep-Disordered Breathing with Positive Airway Pressure Devices: Technology Update." *Medical Devices: Evidence and Research* 8 (2015): 425–37. doi:10.2147/MDER.S70062.

Kimoff, R. J. "When to Suspect Sleep Apnea and What to Do About It." *Canadian Journal of Cardiology* 31, no. 7 (2015): 945–48. doi:10.1016/j.cjca .2015.04.020.

"Sleep Apnoea Trust List of NHS Sleep Clinics in the UK." Last modified July 1, 2014. Sleep Apnoea Trust. http://www.sleep-apnoea-trust.org/sleep-apnoea -trust-list-nhs-sleep-clinics-uk/ (accessed August 25, 2016).

"Sleep Clinic Map." Canadian Sleep Society/Société Canadienne du Sommeil. https://css-scs.ca/resources/clinic-map (accessed August 25, 2016).

"Sleep Clinics." Last modified July 26, 2016. Sleep Disorders Australia. https:// www.sleepoz.org.au/all-clinics (accessed August 30, 2016).

CHAPTER 19: BEATING INSOMNIA WITHOUT PILLS

Cape, J., et al. "Group Cognitive Behavioural Treatment for Insomnia in Primary Care: A Randomized Controlled Trial." *Psychological Medicine* 46, no. 5 (2016): 1015–25. doi:10.1017/S0033291715002561.

Falloon, K., C. R. Elley, A. Fernando 3rd, A. C. Lee, and B. Arroll. "Simplified Sleep Restriction for Insomnia in General Practice: A Randomised Controlled Trial." *British Journal of General Practice* 65, no. 637 (2015): e508–15.

"Find a Sleep Facility Near You [USA]." American Academy of Sleep Medicine. http://www.sleepeducation.org/find-a-facility.

Goldfried, Marvin R., and Gerald C. Davison. *Clinical Behavior Therapy.* Expanded ed. New York: John Wiley and Sons, 1994.

Ho, F. Y., et al. "Self-Help Cognitive-Behavioral Therapy for Insomnia: A Meta-

Analysis of Randomized Controlled Trials." *Sleep Medicine Reviews* 19 (2015): 17–28. doi:10.1016/j.smrv.2014.06.010.

Kabat-Zinn, Jon. *Full Catastrophe Living: Using the Wisdom of Your Body and Mind to Face Stress, Pain, and Illness.* Rev. ed. New York: Bantam, 2013.

Koffel, E. A., et al. "A Meta-Analysis of Group Cognitive Behavioral Therapy for Insomnia." *Sleep Medicine Reviews* 19 (2015): 6–16. doi:10.1016/j.smrv.2014.05.001.

"Let's Build Your Sleep Improvement Program." Sleepio. *http://www.sleepio.com* (accessed August 25, 2016).

Miller, C.B., C. A. Espie, D. R. Epstein, L. Friedman, et al. "The Evidence Base of Sleep Restriction Therapy for Treating Insomnia Disorder." *Sleep Medicine Reviews* 18, no. 5 (2014): 415–24. doi: 10.1016/j.smrv.2014.01.006.

"Online Program for Insomnia." University of Manitoba. http://www.return2sleep .com.

"RESTORE CBT for Insomnia and Sleep." Cobalt Therapeutics. http://cobalttx .com/cbt-online-insomnia-treatment.html (accessed August 25, 2016).

"Sleep Apnoea Trust List of NHS Sleep Clinics in the UK." Last modified July 1, 2014. Sleep Apnoea Trust. http://www.sleep-apnoea-trust.org/sleep-apnoea -trust-list-nhs-sleep-clinics-uk/ (accessed August 25, 2016).

"Sleep Clinic Map." Canadian Sleep Society/Société Canadienne du Sommeil. https://css-scs.ca/resources/clinic-map (accessed August 25, 2016).

Tang, N. K., S. T. Lereya, H. Boulton, M. A. Miller, et al. "Nonpharmacological Treatments of Insomnia for Long-Term Painful Conditions: A Systematic Review and Meta-Analysis of Patient-Reported Outcomes in Randomized Controlled Trials." *Sleep* 38, no. 11 (2015): 1751–64. doi: 10.5665/sleep.5158.

Trauer, J. M., M. Y. Qian, J. S. Doyle, S. M. Rajaratnam, and D. Cunnington. "Cognitive Behavioral Therapy for Chronic Insomnia: A Systematic Review and Meta-Analysis." *Annals of Internal Medicine* 163, no. 3 (2015): 191–204. doi: 10.7326/M14-2841.

CHAPTER 20: MEDICATIONS THAT TREAT SLEEP DISORDERS

"Caffeine Content of Products." Caffeineinformer.com. http://www.caffeine informer.com/the-caffeine-database (accessed August 25, 2016).

Leach, M. J., and A. T. Page. "Herbal Medicine for Insomnia: A Systematic Review and Meta-Analysis." *Sleep Medicine Reviews* 24 (2015): 1–12. doi:10.1016/j .smrv.2014.12.003.

Mendelson, Wallace, and Andrew D. Krystal. "Pharmacology." In *Atlas of Clinical Sleep Medicine*, 2nd ed., ed. Meir H. Kryger, Alon Y. Avidan, and Richard B. Berry, 105–16. Philadelphia: Elsevier/Saunders, 2014.

Newmaster, Steven G., Meghan Grguric, Dhivya Shanmughanandhan, Sathish-
    kumar Ramalingam, and Subramanyam Ragupathy. "DNA Barcoding De-
    tects Contamination and Substitution in North American Herbal Products."
    BMC Medicine 11, no. 222 (2013): 222–35. doi: 10.1186/1741-7015-11-222.
Roth, Thomas. "Pharmacology Section." In *Principles and Practice of Sleep Med-
    icine*, 6th ed., ed. Meir H. Kryger, Thomas Roth and William C. Dement,
    424–505. Philadelphia: Elsevier, 2017.
"Sleep Disorder (Sedative-Hypnotic) Drug Information." Last modified July 8, 2015.
    U.S. Food and Drug Administration. http://www.fda.gov/Drugs/DrugSafety/
    PostmarketDrugSafetyInformationforPatientsandProviders/ucm101557.htm.

ILLUSTRATION CREDITS

CHAPTER 1

Chapter-opening image: Sleeping flamingo.

Brain waves. Adapted from Peter Hauri, *The Sleep Disorders*, 2nd ed. (Kalamazoo, Mich.: Upjohn, 1982).

Brain Centers That Control Sleeping and Waking Functions. Adapted from Meir H. Kryger, Russell Rosenberg, Douglas Kirsh, and Lawrence Martin, *Kryger's Sleep Medicine Review: A Problem-Oriented Approach*, 2nd ed. (Philadelphia: Elsevier, 2015), 20.

CHAPTER 2

Chapter-opening image: John Everett Millais, *L'Enfant du Regiment*. Yale Center for British Art, New Haven, Conn.

CHAPTER 3

Chapter-opening image: Laurent Delvaux, *Ariadne*. Yale Center for British Art, New Haven, Conn.

CHAPTER 4

Chapter-opening image: Christian Krohg, *Mother and Child*. Bergen (Norway) Art Gallery.

CHAPTER 5

Chapter-opening image: Gustav Vigeland, *Couple in Bed*. Gustav Vigeland Studio, Oslo.

CHAPTER 6

Chapter-opening image: Thomas Rowlandson, *The Club Room*. Yale Center for British Art, New Haven, Conn.

CHAPTER 7

Chapter-opening image: Henri Fuseli, *Lady Macbeth*. Musée du Louvre, Paris.

CHAPTER 8

Chapter-opening image: Pierre Puvis de Chavannes, *Le Sommeil*. Metropolitan Museum of Art, New York.

CHAPTER 9

Chapter-opening image: Communications network map of the world. © Maxger/iStockphoto.

CHAPTER 10

Chapter-opening image: Edvard Munch, *Girl Sitting on a Bed*. Bergen [Norway] Art Gallery.

CHAPTER 11

Chapter-opening image: Eugène Delacroix (after Peter Paul Rubens), *Study of Arms and Legs*. Art Institute of Chicago.

CHAPTER 12

Chapter-opening image: Paul Sandby, *Francis Grose Asleep in a Chair*. Yale Center for British Art, New Haven, Conn.

CHAPTER 13

Chapter-opening image: Albert Joseph Moore, *Beads*. Yale Center for British Art, New Haven, Conn.

CHAPTER 14

Chapter-opening image: Vincent van Gogh, *Starry Night*. Museum of Modern Art, New York.

CHAPTER 15

Chapter-opening image: Gustave Courbet, *Sleeping Spinner*. Metropolitan Museum of Art, New York.

CHAPTER 16

Chapter-opening image: Vincent van Gogh, *Bedroom in Arles*. Van Gogh Museum, Amsterdam.

CHAPTER 17

Chapter-opening image: Edgar Degas, *The Absinthe Drinker*. Musée d'Orsay, Paris.

CHAPTER 18

Chapter-opening image: Jacob Jordaens, *As the Old Sing, So the Young Pipe*. National Gallery of Canada, Ottawa.

CHAPTER 19

Chapter-opening image: Alfred Elmore, *Pompeii, A.D. 79*. Yale Center for British Art, New Haven, Conn.

CHAPTER 20

Chapter-opening image: Auguste Rodin, *Le Sommeil*. Rodin Museum, Paris.

CHAPTER 21

Chapter-opening image: Johan Niclas Byström, *Hera with Infant Hercules* (detail). Royal Palace, Stockholm.

INDEX

abdominal pain, 86
Abilify (aripiprazole), 226
abortion, 219
acid reflux, 41, 290
AcipHex (rabeprazole), 206
acne, 36
acromegaly, 209–10
acupuncture, 58
Adderall, 179–80, 286
adenoids, 89, 154, 157, 170
adenosine, 11–12, 284
adolescence, 22–24, 52, 176, 218
Advil (ibuprofen; Motrin), 34, 239
age, changing sleep needs by, 9–10,
    18–28. *See also* older adults
agomelatine, 227–28
AIDS, 223
airline pilots, 6, 107, 109, 110, 117, 118
alarm clocks, 3, 92–93, 130
alcohol: abuse of, 241–42; heartburn
    linked to, 44; jet lag aggravated by,
    99, 100; during pregnancy, 43; RBD
    linked to abuse of, 189; RLS linked
    to, 147; shift work and, 118; SIDS
    linked to parents' use of, 87; sleep
    apnea aggravated by, 62, 159, 165;
    sleep disrupted by, 34, 234, 240–41,
    262, 263; sleepwalking linked to, 84,
    191, 192; snoring aggravated by, 80–81,
    152–53, 160; teeth grinding linked to,
    194

Alertec (modafinil; Provigil), 63, 117, 168,
    178–79, 180, 201, 286, 287
alertness-promoting drugs, 286–87
Aleve (naproxen), 34, 239
Allegra (fexofenadine), 238
allergies, 127, 151, 153, 238, 280
all-nighters, 24, 111–14
alpha-2 agonists, 237
Alzheimer's disease, 56, 75, 90, 198–99
Ambien (zolpidem; Intermezzo), 278,
    279
American Academy of Sleep Medicine
    (AASM), 249, 250, 253
amitriptyline, 222, 280
amphetamines, 179–80, 241, 242, 285,
    286, 287
Anafranil (clomipramine), 232
anastrozole (Arimidex), 240
androgens (male sex hormones), 35–36,
    37, 50, 64
andropause, 63–64
anemia, 43, 64, 140, 141, 142, 143, 145, 147
anger, 33, 35
angina, 202
angiotensin-converting enzyme (ACE)
    inhibitors, 237
angiotensin II receptor blockers, 237
ankles, swollen, 46
antacids, 44, 206
anterior tibialis muscle, 144
anticoagulants, 223

growth hormone, 10
guanfacine (Estulic; Intuniv; Tenex), 286

H1N1 virus (swine flu), 174
hair loss, 63, 208, 210
Halcion (triazolam), 277, 278
hallucinations: in Alzheimer's patients,
    198; hypnagogic, 74, 172, 173, 175,
    178, 187–88; hypnopompic, 187;
    in migraine headaches, 200; in
    narcolepsy patients, 74, 173, 175, 178,
    187, 188; in PTSD patients, 291; in
    RBD patients, 201; in schizophrenia
    patients, 228, 229
Harrison, Benjamin, 111
headaches, 15, 37–38, 58, 221, 224,
    261; drugs used to treat, 240; in pre-
    eclampsia patients, 46; premen-
    strual, 33; in sleep apnea patients,
    158, 160, 199; types of, 199–200
head banging, 75, 194
health insurance, 249, 251–52, 260, 279
heart attacks, 56, 164, 168, 203, 242, 257
heartburn, 15, 41, 44, 160, 205, 206
heart disease, x, 4, 26, 27, 36, 140, 158;
    postmenopausal risk of, 56, 289
heart failure, 158, 164, 197, 203–4, 210,
    213, 257, 290
heart rate, 8, 14, 15, 144, 156; ampheta-
    mines and, 180; blood pressure linked
    to, 158; circadian pattern of, 12; of dia-
    betics, 207; hyperthyroidic elevation
    of, 33, 209; during REM sleep, 186;
    sympathetic nervous system linked
    to, 179
Helicobacter pylori, 206
hemoglobin, 143–44, 147
herbal remedies, 43, 58, 223, 276, 280–81,
    291
heroin, 242
Hetlioz (tasimelteon), 282–83
high blood pressure. See hypertension
hirsutism, 30, 36

histamine, 237–38, 280
hopelessness, 35, 221
Horizant (gabapentin; Neurontin), 58,
    147, 226
hormone production disorders, 207–10
hormone replacement therapy (HRT),
    56–58, 60, 63, 65
hot flashes, 125; in Asian women, 59, 60;
    in breast cancer patients, 63, 212, 240;
    drugs used to treat, 57, 58, 61; in men,
    64; menopausal, 50, 52–53, 54, 55,
    219, 289; premenstrual, 33. See also
    night sweats
hunger, 13
hyperarousal, 124
Hypericum perforatum (Saint John's
    wort), 223
hypertension, 27, 109; defined, 204; drugs
    used to treat, 58, 187, 205, 232, 237,
    286; in PCOS patients, 36; postmeno-
    pausal risk of, 56; sleep apnea linked
    to, x, 46, 164, 169, 204–5
hyperthyroidism, 194, 209
hyperventilation, 269
hypnagogic hallucination, 74, 172, 173,
    175, 178, 187–88
hypnopompic hallucination, 187
hypnosis, 270–71
hypnotics (prescription drugs), 27, 55,
    180, 182, 234, 235, 241, 276, 277–79,
    291
hypocretin (orexin), 174, 175
hypoglycemia, 47
hypothalmus, 31, 52, 53, 210
hypothyroidism, 33, 208–9
hysterectomy, 57

ibuprofen (Advil; Motrin; Nuprin), 34,
    239
idiopathic hypersomnia, 178
imagery-induced relaxation, 270
imipramine, 193, 222, 280
impotence, 64

loratadine (Claritin), 238
losartan, 237
Losec (omeprazole; Prilosec), 206, 238
LSD (lysergic acid diethylamide), 242
lunar cycle, 15
Lunesta (eszopiclone), 278
lung cancer, 109, 212
luteal phase, 31, 32
luteinizing hormone (LH), 56
Luvox (fluvoxamine), 222, 232
lymphoma, 108, 211
Lyrica (pregabalin), 211

MacNish, Robert, 6
Macrobius, Ambrosius Aurelius Theo-
    dosius, 9
mad cow disease (bovine spongiform
    encephalopathy), 134
Mairan, Jean-Jacques d'Ortous, 13–14
mammals, 14, 173
mammogram, 56
manic depressive (bipolar) disorder, 35,
    197, 199, 217, 224–26
MAOIs. See monoamine oxidase inhib-
    itors
marijuana, 213, 280, 281
marine animals, 191
massage, 45, 213
mastectomy, 63
mattresses, 127–28
mazindol (Mazanor; Sanorex), 286
MDMA (Ecstasy), 242
meat, 146
medical workers, 6, 110, 113, 118
Medicare, 251–52
meditation, 271–72
melanopsin, 227
melatonin, 12, 13, 228; cancer linked to
    reduction in, 109; electronic devices
    and, 22; as medicinal supplement, 96,
    97, 100, 189, 199, 227, 237, 281–82
memory loss, 33, 99, 100, 133, 198, 211,
    224

menarche, 124
menopause, 17, 27, 33, 49, 50, 139, 284;
    in cancer patients, 51–52, 63, 212,
    240; effects of, 53–54, 194, 205, 289;
    insomnia during, 124; mood changes
    linked to, 219; sleep problems after,
    61–62; sleep problems during, 54–55,
    170; symptoms of, 52; treating sleep
    problems from, 55–61
menstrual cycle, 14–15, 17, 29–32, 50,
    62, 108; mood changes during, 219;
    PCOS and, 36; phases of, 31; PMS
    and, 32–26; RLS linked to, 142, 146
mental disorders, 215–33, 291. See also
    specific disorders
metabolic disorders, 36, 47, 109
metformin, 37
methadone, 239
methamphetamine, 285, 286
methyldopa, 237
methylphenidate (Concerta; Ritalin),
    176, 179, 242, 285–86
Middleton, Kate, duchess of Cambridge,
    41
Midol, 34
migraines, 199–200
milk, 44, 86
milnacipran HCl (Savella), 211
Mindell, Jodi, 85
mindfulness, 271–72
Mirapex (pramipexole), 141, 147
miscarriages, 40, 43, 47, 285
miso, 59
mitrazipine (Remeron), 222
modafinil (Alertec; Provigil), 63, 117, 168,
    178–79, 180, 201, 212, 286, 287
monoamine oxidase inhibitors (MAOIs),
    222, 223
mood swings, 33, 35, 48, 54, 61, 225
morning sickness, 41
morphine, 239
Motrin (ibuprofen), 34, 239
mouth breathing, 151